# TEACHING MATHEMATICAL THINKING

*Tasks and Questions to Strengthen Practices and Processes*

• • • • •

# TEACHING MATHEMATICAL THINKING

*Tasks and Questions to Strengthen Practices and Processes*

• • • • •

**Marian Small**

*Foreword by* Linda Dacey

TEACHERS COLLEGE PRESS
TEACHERS COLLEGE | COLUMBIA UNIVERSITY
NEW YORK AND LONDON

NELSON
nelson.com

Published by Teachers College Press, 1234 Amsterdam Avenue, New York, NY 10027; distributed in Canada by Nelson Education, 1120 Birchmount Road, Toronto, ON, Canada M1K 5G4.



*Text Design:* Lynne Frost

*Library of Congress Cataloging-in-Publication Data*

Names: Small, Marian.
Title: Teaching mathematical thinking : tasks and questions to strengthen practices and processes / Marian Small.
Description: New York, NY : Teachers College Press, [2017] | Includes bibliographical references and index.
Identifiers: LCCN 2016055513 (print) | LCCN 2016059745 (ebook) | ISBN 9780807758564 (pbk. : alk. paper) | ISBN 9780807775882 (ebook)
Subjects: LCSH: Mathematics teachers—Training of—Canada. | Mathematics teachers—Training of—United States. | Mathematics—Study and teaching (Elementary)—Canada. | Mathematics—Study and teaching (Elementary)—United States. | Mathematics—Study and teaching (Middle school)—Canada. | Mathematics—Study and teaching (Middle school)—United States.
Classification: LCC QA10.5 .S63 2017 (print) | LCC QA10.5 (ebook) | DDC 372.70973—dc23
LC record available at https://lccn.loc.gov/2016055513

ISBN 978-0-8077-5856-4 (paper)
ISBN 978-0-8077-7588-2 (ebook)

Printed on acid-free paper
Manufactured in the United States of America

25 24 23 22 21 19 18 17 8 7 6 5 4 3 2 1

• • •

# Contents

• • •

# Foreword

**WHEN ASKED** to write this Foreword for Marian Small's *Teaching Mathematical Thinking: Tasks and Questions to Strengthen Practices and Processes,* I was excited. It meant I would have an early opportunity to read the book. Many years ago, Marian and I worked together on a writing project for the National Council of Teachers of Mathematics, and I have followed Marian's work ever since. Whether listening to a five-minute Ignite talk or reading one of her books, I am always captivated. To have a new book—with this great title—my expectations were high. With movies and books, such anticipation can result in my being disappointed. Not so here.

In the United States, the Standards for Mathematical Practice describe proficiencies we want all our students to gain. In Canada, process standards are articulated, which have many similarities with the practice standards, although more attention is given to visualization, mental math, and estimation. I have heard many arguments against particular content goals but never disagreement with these practice and process standards. Yet several years after their release, many uncertainties remain about what they actually mean and how they can be achieved.

In my work with teachers, I often hear questions such as *Where can I find problems that would require students to persevere? What does justification look like in a second-grade classroom? What's the difference between looking for structure and looking for regularity?* This book offers us the insights we need to more deeply understand the practice and process standards. It also helps us understand how to facilitate students' development of these proficiencies, that is, how to help them become mathematical thinkers. The book includes a chapter for each practice standard and then a final chapter on visualization, mental math, and estimation. The last chapter should not be thought of as an add-on; visual thinking permeates the book, and mental math and estimation are highlighted throughout as appropriate.

Each chapter gives attention to how these standards work across three grade spans (K–2, 3–5, and 6–8), allowing the reader to more fully understand each standard as well as how its application develops over time. Tasks and problems—along with variations, possible responses, follow-up questions, and examples of student work—weave a vision of teaching for mathematical thinking. Chapters end with a thoughtful "look for" list to access student proficiency, as well as a summary of key ideas for engaging students.

Readers of Marian's earlier book, *Good Questions: Great Ways to Differentiate Mathematics Instruction* (2nd edition, 2012), will not be surprised by the quality of each chapter's tasks and problems. Much of my career has focused on problem solving, and Marian's work never ceases to provide me with new ideas. In this book, my favorite example of her ingenuity is a problem she suggests on page 110. She takes the familiar and worthwhile task of showing students a set of dots and asking about the different ways to count them and transforms it, by making a simple twist. She poses, *How would you arrange 8 dots to make it easy to quickly see it as 8?* This is such a modest change, and yet it alters the way the students must think, requiring them to create their own visual representation rather than respond to one provided.

As important as the quality of the tasks we present is, we also need to support students' productive struggle as well as further challenge their thinking. It is here that Marian's discussions of how students might respond and what questions we might ask are invaluable, offering teachers the tools needed to actualize these standards. These sections of the book also further highlight important mathematical ideas.

Whether you are a new teacher or a seasoned educator, this book will enrich your ability to develop your students' mathematical thinking. (I suspect you'll find that it develops your own mathematical thinking as well.) While you will gain much from reading it alone, you may want to find a colleague to read it with you, or form a professional learning community. A collaborative investigation will further enrich your consideration of these ideas.

I appreciate the way Marian Small has provided a clear understanding of each standard and of how they interrelate. I am grateful for her suggestions for follow-up questions that will help us further probe students' thinking. But my favorite part of this book is the rich tasks and problems she poses to help readers better support these practices and processes. I predict you will be intrigued by them as well and will want to engage in "mathematical play" as soon as you read them. Keep a pencil and paper nearby, and plan to return this book many times, over many years, and enjoy!

—Linda Dacey
*Professor Emerita, Mathematics*
*Lesley University*

• • •

# Preface

## ORGANIZATION OF THE BOOK

This book begins with an Introduction that describes the history of the Mathematical Standards for Practice in the Common Core curriculum (Common Core State Standards Initiative, 2010) and the process standards addressed in Canadian curricula, followed by eight chapters that focus on each of the eight standards of practice and a ninth chapter that addresses the processes of visualization and mental math and estimation. I believe that inclusion of the Canadian process standards may provide additional support to U.S. teachers in their instructional planning.

Each chapter includes some general discussion about how to ensure that the particular standard or process being considered receives appropriate attention, as well as a discussion of what each standard of practice might look like in Grades K–2, Grades 3–5, and Grades 6–8. Many of the problems are rich enough to be valuable in more than one of these grade groups, and this is indicated when that is the case. Within the discussion of the problem, variations on how to use the problem for different grade bands and expectations of ways in which students might respond are described.

Many examples of student work are provided, both to clarify the ideas being raised and to provide suggestions for how feedback might be offered when the illustrated problems are used with other students.

## ACKNOWLEDGMENTS

This book has been developed thanks to the continued confidence shown in me by Teachers College Press. It was their suggestion that we create this book. Putting this material together has been rewarding because of the importance of helping teachers work out ways to highlight the standards of practice.

To support my many Canadian colleagues, I have chosen to include material focused on the processes of the mathematics curricula used in Canada that are not directly addressed by the Common Core Mathematical Standards of Practice.

Lastly, I thank a number of my consultant/coach colleagues who have helped me gather the student samples used in this work.

# TEACHING
# MATHEMATICAL THINKING

*Tasks and Questions to Strengthen Practices and Processes*

• • • • •

• • •

# Introduction

## BACKGROUND

Most of us, when we think about math, think of what is generally considered *content*. We talk about working with fractions, or understanding place value, or solving equations. But in the background there have always been the fundamental mathematical practices and processes. Concepts such as reasoning, problem solving, recognizing structure, and modeling are not new to us.

One of the big changes instituted in the most recent curriculum revisions both in the United States and in Canada has been to bring these background processes to the foreground, not only to highlight their importance but also to help us see that if we focus explicitly on these processes, then content will be learned differently and may in fact be learned more effectively.

Much of this change began with the 1989 National Council of Teachers of Mathematics (NCTM) *Curriculum and Evaluation Standards for School Mathematics* (NCTM, 1989). In this document, notions of math as problem solving, as communication, and as reasoning and mathematical connections were explicitly discussed.

In 2000, NCTM updated the standards in the *Principles and Standards for School Mathematics* (NCTM, 2000) and specifically mentioned the following process standards:

- Problem solving
- Communication
- Representation
- Reasoning and proof
- Connections

These standards served as an underpinning for many state and provincial curricula.

More recently, the advent of the Common Core Standards in the United States has evolved these process standards into explicitly stated Standards for Mathematical

Practice. Canadian curricula still focus on process standards, although the names of the processes vary across the country.

The eight Standards for Mathematical Practice from the Common Core curriculum are listed here. They will be covered in detail in subsequent chapters.

## COMMON CORE STANDARDS FOR MATHEMATICAL PRACTICE

1. Make sense of problems and persevere in solving them. (Related to the processes of problem solving and reflecting [one of the Canadian process standards listed below])
2. Reason abstractly and quantitatively. (Related to the processes of reasoning and representation)
3. Construct viable arguments and critique the reasoning of others. (Related to the processes of reasoning and communication)
4. Model with mathematics. (Related to the process of representation)
5. Use appropriate tools strategically. (Related to the process of representation)
6. Attend to precision. (Related to the process of communication)
7. Look for and make use of structure. (Related to the process of connections)
8. Look for and express regularity in repeated reasoning. (Related to the process of connections)

Many rich mathematical activities involve more than one of these standards at the same time, as will be noted in the later chapters.

## CANADIAN CURRICULA PROCESS STANDARDS

In different provinces in Canada, these process standards are articulated in various curricula (Ministère de l'Éducation, Gouvernement du Québec, 2001; Ministry of Education Ontario, 2005; Western and Northern Canadian Protocol, 2006).

- Problem solving (or situational problem solving)
- Communication
- Representing
- Reflecting
- Technology
- Mental math and estimation
- Reasoning
- Connecting or Connections
- Selecting tools and strategies
- Visualization

The processes of problem solving, reasoning, representing, and connecting are directly linked to the Common Core Standards for Mathematical Practice.

The process of reflecting is an integral part of Mathematical Practice Standard 1, when students reflect on solutions, although reflecting could also include reflection on strategies, connections, or other aspects of mathematical work as well.

The processes of visualization and mental math and estimation are less directly addressed in the Common Core Standards for Mathematical Practice. Visualization is often embedded in problem solving (Practice Standard 1), constructing viable arguments (Practice Standard 3), and modeling (Practice Standard 4), but not necessarily. Mental math and estimation is often linked to looking for and making use of structure (Practice Standard 7), but not necessarily. Therefore, I have included an additional chapter in this resource (Chapter 9) to focus specifically on the processes of visualization and mental math and estimation.

## MISCONCEPTIONS ABOUT THE STANDARDS FOR MATHEMATICAL PRACTICE AND THE MATHEMATICAL PROCESSES

Because of overall concerns with the Common Core, but particularly because the Standards for Mathematical Practice and the mathematical processes are "separated" from the other standards or outcomes/expectations in various curricula, many misconceptions can get in the way of using them as intended.

Some teachers believe that the intention is that all lessons must incorporate all standards for practice or all processes (Mateas, 2016). This is certainly not the case. Few, if any, lessons would include all of them, although many might include more than one. Although each standard and several processes are described separately in the ensuing chapters, I often reference how different standards can apply while working on the same problem. I also give a great deal of attention to when a particular standard or process might be appropriate.

Some believe it is possible to predetermine which standard or process will arise when students are confronted with a particular task. Although it is true that some tasks are more likely to elicit particular standards or processes than others, it is always up to the student what she or he brings to bear when working on a problem or in a mathematical situation.

Some teachers believe that the standards and processes are not taught; they just happen. Indeed, for some students, they will just happen. But there is definite benefit in articulating explicitly what these standards and processes are. This can only help students become more aware of their thinking and extend their thinking in new situations.

Some teachers believe that the standards or processes are not taught at the same time content is taught. This is not likely in that the descriptions of how to apply a standard or process require mathematical content in which to be embedded.

## PROMOTING MATHEMATICAL THINKING

An overarching goal of teaching and making students aware of mathematical practices and processes has been to move students away from simply using algorithms for mathematical computation and to move them toward becoming mathematical thinkers. To this end, teachers have stepped back from simply telling students what to do and giving them practice at doing it. Instead they allow students to sometimes stumble as they move forward.

However, there is a fine line to be considered. I occasionally encounter teachers who feel that they should never give students any assistance. In fact, it is important for teachers to be attuned to when struggle is productive and when students need encouragement and prompts to their thinking in order to move forward. For this reason, I offer many descriptions in this resource of what questions to ask or what actions to take when students stumble.

• CHAPTER 1 •

# Making Sense of Problems and Persevering in Solving Them

## *Mathematical Practice Standard 1*

***MP1. Make sense of problems and persevere in solving them.***

Mathematically proficient students start by explaining to themselves the meaning of the problem and looking for entry points to its solution. They analyze givens, constraints, relationships, and goals. They make conjectures about the form and the meaning of the solution and plan a solution pathway rather than simply jumping into a solution attempt. They consider analogous problems, and try special cases and simpler forms of the original problem in order to gain insight into its solution. They monitor and evaluate their progress and change course if necessary. Older students might, depending on the context of the problem, transform algebraic expressions or change the viewing window on their graphing calculator to get the information they need. Mathematically proficient students can explain correspondences between equations, verbal descriptions, tables, and graphs or draw diagrams of important features and relationships, graph data, and search for regularity or trends. Younger students might rely on using concrete objects or pictures to help conceptualize and solve a problem. Mathematically proficient students check their answers to problems using a different method, and they continually ask themselves, "Does this make sense?" They can understand the approaches of others to solving complex problems and identify correspondences between different approaches.

**THIS STANDARD** builds on a long-standing belief that problem solving is a critical part of mathematics (Polya, 1957) and a belief that students must reflect on their problem-solving behaviors. It suggests, in particular, that students must not only make sense of the problems they are solving, but they must also be persistent enough to arrive at a solution.

Students can be coached to adopt the problem-solving methodologies described above and given sufficient opportunities to consciously use them so that selecting and applying these methodologies becomes second nature for them.

## WHAT IS A PROBLEM?

A problem is a puzzle. There is no immediate certainty about how to proceed to solve it, although it may be obvious that it has to do with mathematics. That is why it is so important that, for example, not all problems in a multiplication unit are about multiplication; if they were, how to proceed would be much more immediately obvious and the problems might not really be problems.

Problems need not be word problems, but they could be. Examples of word problems and non–word problems will be included in this chapter. As well, examples of open-ended problems, problems with multiple answers, and problems with many possible approaches but only one solution will be included.

## WHAT IS INVOLVED IN INTERPRETING A PROBLEM?

How students make sense of problems is a complex process. Students must recognize that they first have to make sense of what is being asked in order to effectively solve a problem. This involves the following:

- Determining the givens or what is known in a problem
- Recognizing constraints or restrictions that must be considered
- Looking for relationships among the givens
- Becoming aware of what is required in a solution

## THE NEED FOR A PLAN

Because problems are puzzles, it is critical that students make plans for solving them. Learning how to make the plan should be a critical piece of mathematics instruction.

Many years ago, Polya (1957) suggested that there were four key steps in solving a problem:

*Stage 1: Understand the problem*
*Stage 2: Make a plan*
*Stage 3: Carry out the plan*
*Stage 4: Look back*

Stage 4 involves looking at whether an answer is reasonable or makes sense. The importance of this step must not be underemphasized.

Since then, there have been a number of variations on Polya's (1957) listing of key steps. Zollman (2009) takes a graphic organizer approach. In Zollman's organizer a center section asks what needs to be found. This is surrounded by four quadrants: what is already known, brainstorming on solution methods, solution attempts, and a listing of explanations needed in a write-up. The diagram below is adapted from Zollman's (2009) organizer:

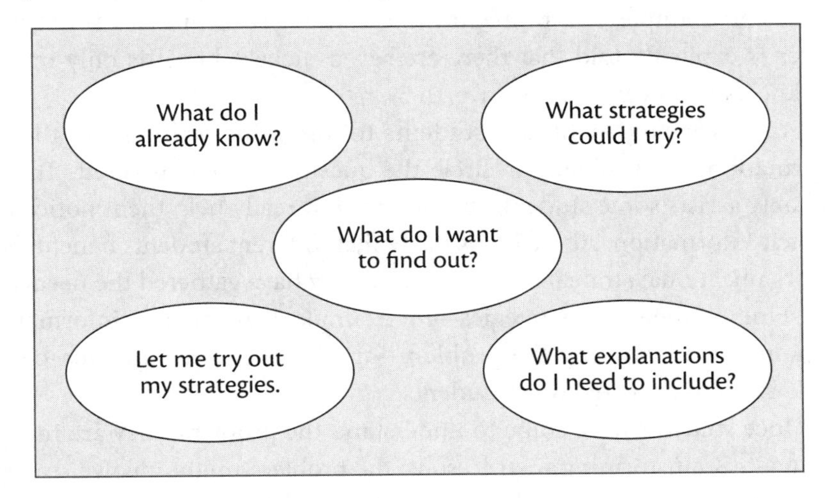


For example, a problem might be: You buy 4 identical items that cost less than $20 each. You buy 8 identical items that cost more than $20 each. In total, you spend about $300. How much could each set of items have cost?

Working from my adaptation of Zollman's (2009) organizer, a student might write:

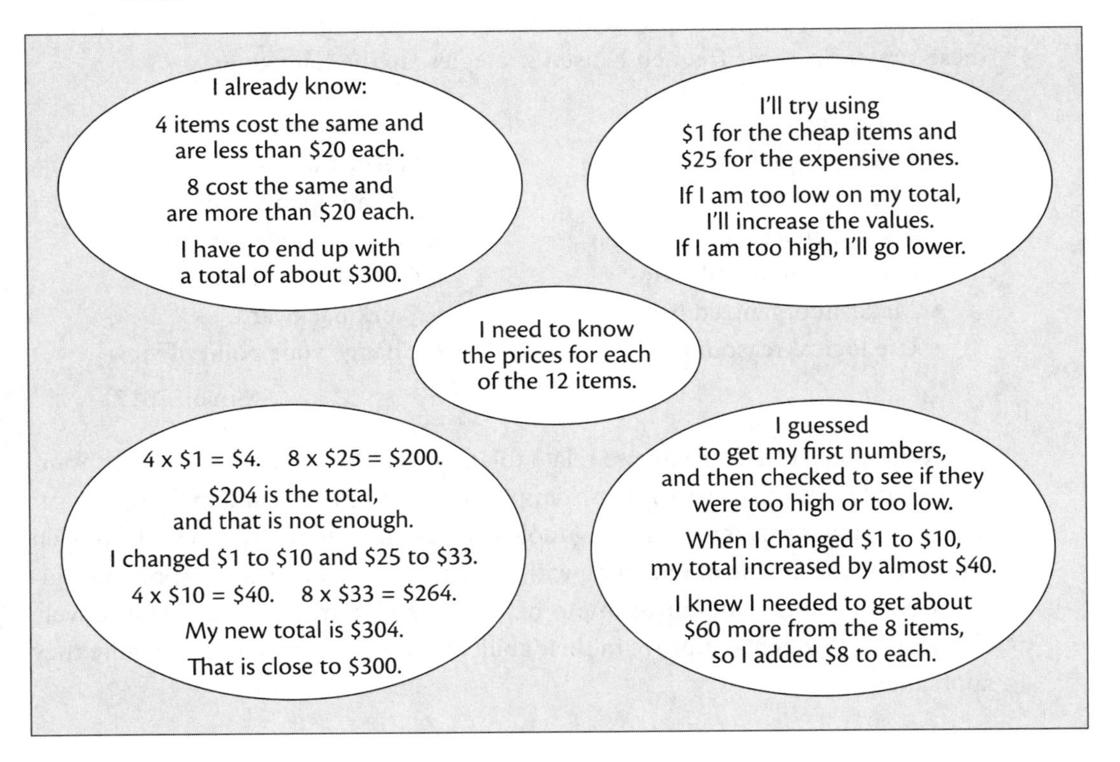

In looking at what they already know when exploring a problem, students need to pay attention to all of the information that is either explicitly or implicitly given. For example, if a problem indicates that someone has seven nickels, the reader is explicitly told that there are seven nickels; but it is only implicitly that one knows that every nickel is worth 5¢.

Some teachers encourage students to underline or highlight the important information in a problem or circle the question to be answered. This approach probably assists some students, but it does not really help them notice the critical implicit information. As well, it is likely that different students benefit from different organizational strategies to make sure they have gathered the needed information. Finding their own strategies—for example, copying over information as bullet points or illustrating the problem—might suit some students better than a single method expected of all students.

Once students have come to understand the problem, they are ready to begin solving it. Brainstorming ways to solve the problem might involve strategies listed in the next section.

## WHAT STRATEGIES MIGHT STUDENTS USE TO SOLVE PROBLEMS?

For many years, mathematics programs have shared sets of problem-solving strategies that apply to many problems. Although not all problems are solved with these strategies, some frequently used strategies are the following:

- Act it out
- Draw a picture
- Look for a pattern
- Make a chart/table or graph
- Consider all possibilities
- Make an organized list
- Use logical reasoning
- Use a model
- Guess and test
- Use an open sentence
- Solve a simpler problem
- Consider extreme cases
- Work backward
- Change your point of view

(Small, 2017)

A number of these strategies relate to additional Mathematical Practice Standards and not just Standard 1. For example, drawing a picture and making a chart/table or graph are examples of using tools strategically (Standard 5). Using an open sentence is an example of modeling with mathematics (Standard 4). Applying logical reasoning might be an example of reasoning abstractly and quantitatively (Standard 2). Students improve in their ability to use these strategies the more they apply them.

## WHAT DO WE NEED TO DO TO DEVELOP PERSEVERANCE IN PROBLEM SOLVING?

Perseverance (or grit) (Duckworth, Peterson, Matthews, & Kelly, 2007) does not always come automatically; it must be nurtured. And it can only be nurtured if students are asked to solve problems that are real problems, not immediate applications of what they have been shown. In addition, the problems need to be at an appropriate level, that is, within the students' zone of proximal development. It is critical that students be able to solve the problem, given their current skill sets, if they persist. If, too often, perseverance does not lead to success, students are unlikely to persevere in future situations.

Another habit related to perseverance that students need to develop is the habit of looking for more than one answer, even after a first one is found. Very often, students expect that there is only one answer and simply stop when they have found it. They need to learn the habit of always wondering if there could be more answers and checking for them. This habit is learned when teachers regularly remind students to look for alternate solutions and frequently offer problems with many solutions.

## EXAMPLES OF PROBLEMS THAT MIGHT BRING OUT MATHEMATICAL PRACTICE STANDARD 1

### Grades K–8

**The Counters Problem**

- I have some counters.
- I split them up into 2 small piles and 2 large piles.
- Each small pile has the same number of counters.
- Each large pile has the same number as the total of the 2 small piles.

? How many counters might there be, in total, in the 4 piles? Find a number of possible totals.

#### *What Kind of Plan Might a Student Make?*

A student needs to realize that

- There are a total of four piles.
- Two piles are big and two are small.

- The big piles are equal in size and the small piles are equal in size.
- A big pile has twice as many counters as a small pile.

He or she needs to figure out different possible total numbers of counters.

A plan might involve guess and test:

Small Pile = 2
Big Pile = 4

The total is 12 counters.

Some students will quit once they achieve one solution and simply say that the answer is 12. It is up to the teacher to intervene and say:

- *Does it have to be 12?*
- *What if your small pile had been a different size?*

Then students might try a different number, for example, 4 counters in a small pile:

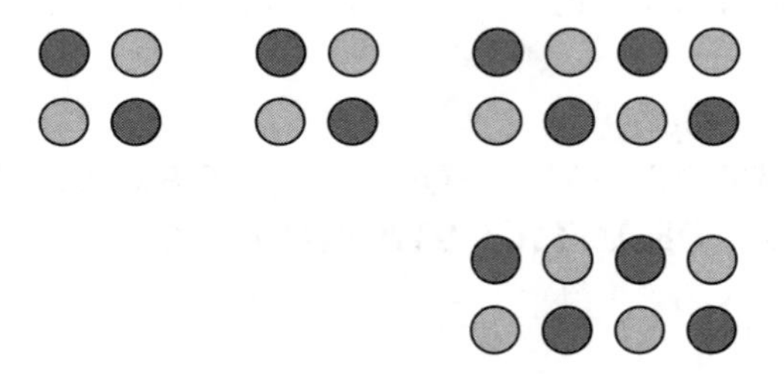

The total this time is 24.

#### *How Does a Teacher Push the Student to Go Farther?*

Many students will now suggest they are done. They would argue that since the answers were 12 and 24, they know that the total number of counters is an even number.

A teacher needs to prompt, for example:

- *Could the total be 4? 4 is an even number.* [By the way, it cannot.]
- *Could it be 20? 20 is an even number.* [Again, it cannot.]

Once students have more possibilities—for example, totals of 12, 24, 18, 60, 30—a teacher needs to ask:

- *What do all of these numbers have in common, besides being even?*

It might be helpful to write some of the numbers in order: 12, 18, 24, 30, . . . . Then students might notice that the answers go up by 6 each time.

Ideally students will use their own initiative to test a conjecture—that the total number of counters can be any number in the sequence 6, 12, 18, 24, 30, 36, 42, . . .—or they might need to be led to test that hypothesis. In the end, this turns out to be correct; the total number of counters is a multiple of 6. This conjecture could be posited and tested, which would be an application of Mathematical Practice Standard 3, constructing and critiquing arguments. Students should test a hypothesis by using a strategy other than the one they began with. They might see that the conjecture makes sense by thinking of the piles like this:

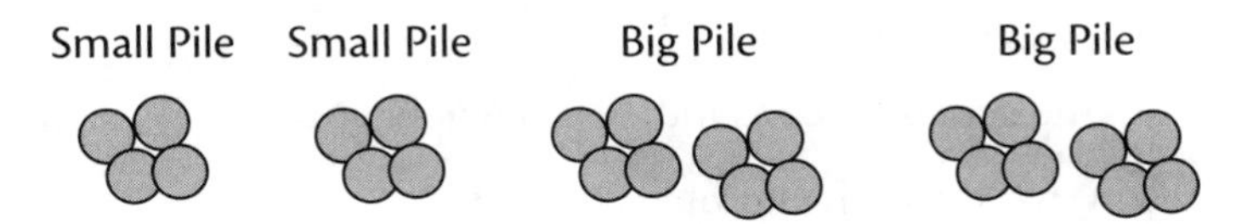


They could see that each time there are essentially 6 groups of however many counters are in a small pile (1 group in the first small pile, 1 in the second, 2 in the first big pile, and 2 in the second big pile), so it makes sense the totals are numbers like 6, 12, 18, 24, . . . .

Note a few things about this problem. First, it might have been approached using a different strategy (not guess and test), for example, making an organized list or creating a table or chart. Older students, as will be described below, might use an open sentence or logical reasoning.

Second, the problem is based on thinking of one number as groups of another, the heart of proportional reasoning, something we know is valuable for students (Dole, Clarke, Wright, & Hilton, 2010).

### *How Does This Problem Build Perseverance?*

This problem helps build perseverance in several ways. First, it has many answers, and students are asked to provide several of them. That alone helps build perseverance since students cannot quit the minute they get something down on paper.

The fact that there are many answers also pushes students to look for relationships. They could ask, for example, *What do these answers have in common?* This seeking of relationships is at the heart of mathematics learning, and looking for relationships is exactly what we need students to persist at.

### *Solutions from Older Students*

Students who are young are likely to test their answers by showing how those numbers are achieved. Some older students solving this same problem might notice that once one answer works, adding 1 more counter to each of the small piles and 2 more to each of the large piles has the effect of adding 6, and that new total will also work.

Students who are in Grades 6–8 might apply Mathematical Practice Standard 4 and make a model. They might represent the small pile value as $x$ and the larger pile value as $2x$. The total number of counters is $x + x + 2x + 2x = 6x$. So the answers are, indeed, always multiples of 6.

**Sums and Differences**

- I add two numbers.
- I also subtract them.
- The sum is twice as much as the difference.

? What could the numbers be?

What could they NOT be?

***What Kind of Plan Might a Student Make?***

A student needs to realize that

- No numbers are given, but he or she has to find them.
- She or he has to both add the numbers and subtract the numbers.
- It is the sum that is double the difference, not one number that is double the other.

The goal is to determine pairs of numbers that work, as well as pairs of numbers that do not work.

At this point, most students will just try some numbers. If a student is lucky and an early pair of numbers works, he or she is likely to persist and keep trying new numbers. But many students will give up after they try even two or three pairs of numbers that do not work. At this point, a teacher might intervene.

***How Does a Teacher Push the Student Forward?***

If a student has found a pair of numbers that meets the required condition, the teacher needs to say:

- *That's great that 2 and 6 work. What do you notice about those numbers?*

When the student says that 6 is 4 more than 2, the teacher could ask:

- *Why don't you try another pair of numbers that are 4 apart and see if they work?*
- *What else is true about 2 and 6?*

If a student has been unsuccessful, a teacher might say:

- *I noticed that when you chose 2 and 4, the sum was three times the difference. If you chose 1 and 4, would you be getting closer to twice the difference or farther away?*

The student whose work is shown below has found a couple of pairs that work:

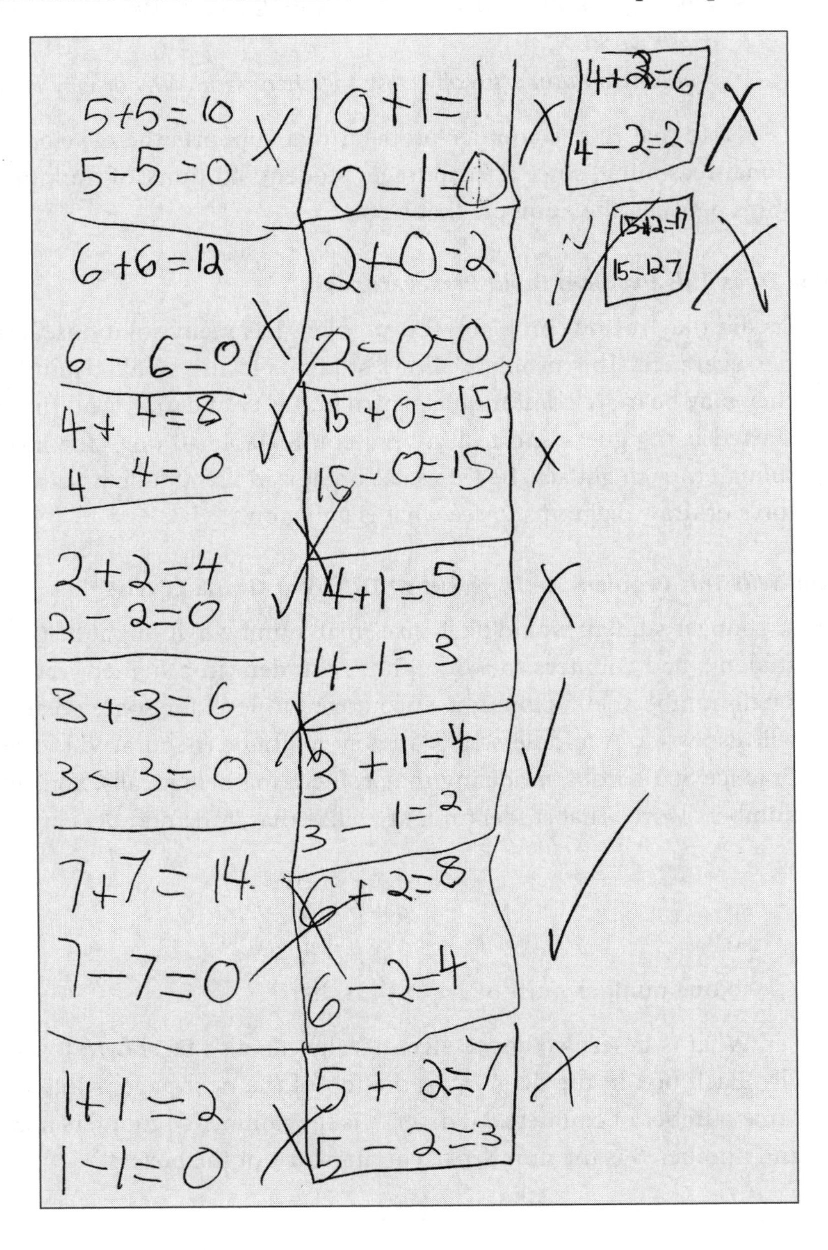

In this case, a teacher might ask:

- *So how are the pairs 1 and 3 and 2 and 6 alike?*
- *Can you draw a picture to show why those pairs make sense?*

If an older student has used only small numbers, a teacher might ask:

- *What big numbers do you think might also work? Why do you think those might work?*
- *Would two numbers really close together work? Why or why not?*

Note that this is another problem that supports the development of proportional reasoning, since it encourages students to think of multiplicative relationships between the numbers that result.

### *How Does This Problem Build Perseverance?*

As did the previous problem, this problem has many solutions, which helps build perseverance. This problem allows students to use smaller numbers with which they may be more comfortable, which builds confidence. And the problem can be started using guess and test, a very comfortable strategy for most students, although they might also be led to use organized lists. Students can also use physical tools or draw diagrams to see what is going on.

### *How Will This Problem Be Different at Different Grade Levels?*

A younger student would likely use small numbers. It might help if those younger students had counters to work with. A student in a higher grade might still use small numbers, but should be encouraged to look for some greater numbers that will also work. A middle-school student might be encouraged to use Mathematical Practice Standard 4, modeling the problem mathematically, and see what sorts of numbers work. That student might realize that if the numbers are $a$ and $b$, then

$$a + b = 2(a - b)$$

so $$a + b = 2a - 2b$$

so $$3b = a,$$

so one number must be triple the other

What is described above algebraically can also be shown to make sense visually. Each box in the diagram at the top of the next page is intended to hold the same number of counters. Number A is the number of counters in 3 of those boxes and number B is the number of counters in 1 of the boxes.

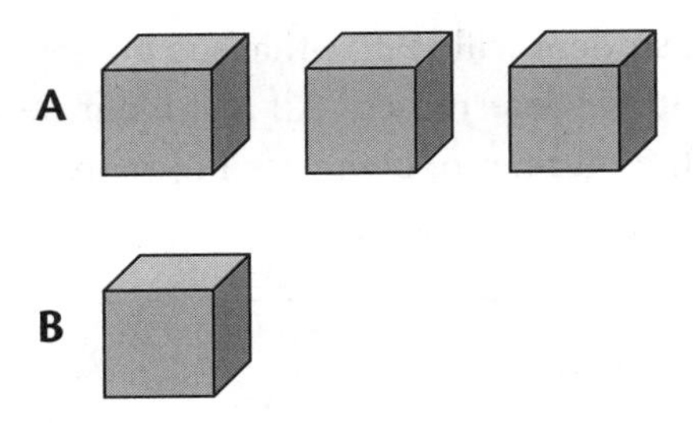


Since all of the boxes hold the same number of counters, the sum of $A$ and $B$ is 4 times the number in 1 box. The difference, which tells how much more $A$ is than $B$, is the number in 2 boxes. And what is in 4 boxes is always double what is in 2 boxes.

## Grades K–2

> **Trading Coins**
>
> - I represent an amount of money with 8 coins.
> - I represent the same amount with 22 coins.
>
> ? What coins might I have had each time?

### *What Kind of Plan Might a Student Make?*

A student needs to realize that

- The two sets of coins represent the same amount.
- There is an increase of 14 coins (implicit information).
- The coins to choose from are pennies, nickels, dimes, and quarters (in the United States) or nickels, dimes, quarters, loonies ($1 coins), and toonies ($2 coins) (in Canada).

The goal is to list the sets of coins for each situation and to provide more than one possibility, if there is one.

To create a plan, some students might simply choose values and try to come up with different numbers of coins to make up that value and hope an answer arises. For example, a student might choose a value like 50¢ and try to represent it different ways, hoping that one way uses 8 coins and one uses 22 coins. Sometimes, this works. Unfortunately, this will often prove not to be a useful strategy; for example, it won't work with 50¢. At this point, a teacher should probably ask students to talk to other students who might be trying a different approach. This provides an opportunity for students to see each other as resources.

Hopefully, a student will realize that the total value of the groups of coins does not really need to be determined. All a student needs to consider is how a coin trade increases the number of coins used to represent a value.

Here is one student's thinking:

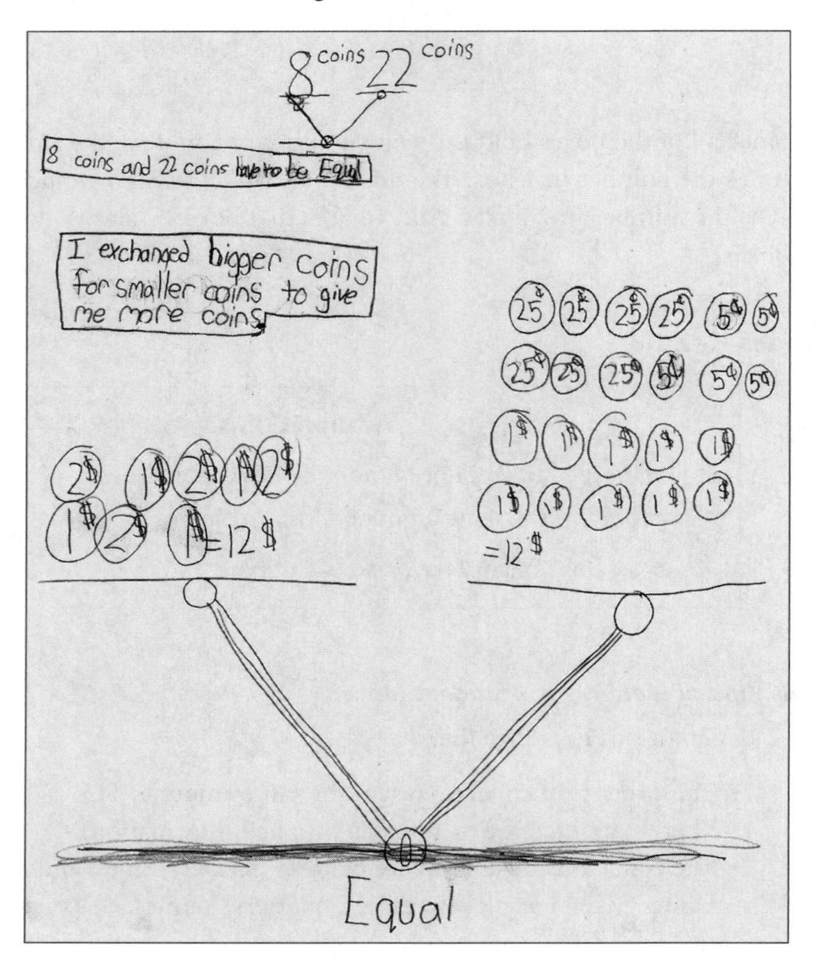


The student whose work is shown above tells us that she realized she had to trade more valuable coins for less valuable coins, which is good information to share in a solution, but she does not tell us anything about how she made the choices she did. It looks like she traded one $2 coin for 7 quarters and 5 nickels, an increase of 11 coins, and that she traded the other three $2 coins for six $1 coins, an increase of 3 more coins, but she does not tell us why she chose $12 to work with or how she figured out which coins to trade.

Similarly, the student whose work is shown on the next page traded quarters for dimes and nickels to get more coins, but doesn't tell us how she chose the values she did.

***How Does a Teacher Push the Student Forward?***

A teacher should be pleased with each of these students' work, but both students need more instruction in or attention to what is required to communicate mathematically.

A student might have listed all the possible coin trades. For example:

1 nickel = 5 pennies (an increase of 4 coins)
1 dime = 10 pennies (an increase of 9 coins)
1 dime = 2 nickels (an increase of 1 coin)
1 quarter = 2 dimes and a nickel (an increase of 2 coins)
1 quarter = 5 nickels (an increase of 4 coins)
1 quarter = 1 dime and 3 nickels (an increase of 3 coins)
1 loonie ($1 coin) = 4 quarters (an increase of 3 coins)
1 loonie = 10 dimes (an increase of 9 coins)
1 loonie = 5 dimes and 10 nickels (an increase of 14 coins)
1 toonie ($2 coin) = 2 loonies (an increase of 1 coin)
and so forth

The teacher could then prompt with one or more additional questions, such as the following:

- *Could you have had 8 nickels and traded one of those nickels for pennies? Why not?*
- *How many extra coins do you need to get?*
- *Does anything on your list help you get there?*
- *Do you have to get 14 more coins based on just a single trade?*

A student might check his or her work by determining the values of the coins in each situation or might test answers, instead, by matching up sub-amounts of equal values (e.g., quarters with quarters or a quarter with 2 dimes and a nickel) from each combination.

A teacher might ask students why a particular solution makes sense and hope they realize that if there are more coins, those coins have to be of lower value. This is a fundamental concept in the building of proportional reasoning.

#### *How Does This Problem Build Perseverance?*

This problem has many solutions, which helps build perseverance. It uses a familiar context, which builds confidence and, therefore, perseverance. And the problem can be acted out using guess and test, a very comfortable strategy for most students.

While some students would need the coins in front of them in order to actually perform the trades, other students are likely to be able to be somewhat more abstract.

For Canadian students, noticing that a loonie (a $1 coin) could be traded for 5 dimes and 10 nickels leads to a very simple answer. Start with 8 loonies and trade only one for 5 dimes and 10 nickels. That alone increases the total number of coins from 8 to 22.

But there are many other possible answers, for either U.S. or Canadian students. For example, since $14 = 4 + 4 + 4 + 2$, a student could look for a situation where there is an increase of 4 coins, to be used three times, and a situation where there is an increase of 2 coins, to be used once.

For example, if there had been 7 quarters and a dime (for a total of $1.85), a student might trade 3 quarters for 5 nickels each, increasing the number of coins by 12, and another quarter for 2 dimes and a nickel, increasing the coin count by 2 more, for a total increase of 14, bringing the final count to 22 coins.

Since 14 is seven 2s, students could also begin with 8 quarters and trade 7 of them for 2 dimes and 1 nickel each (adding 2 coins each time), leaving the other quarter alone.

There are many additional answers, and students should be encouraged to look for many of them. Students could be encouraged to check some of the answers by totaling the monetary amounts.

➢ ***Variations.*** This problem is an example of a group of problems that could be created involving coin or bill trades. For example, a student might be told that an amount could be represented with either 4 coins or 7 coins, or with either 2 bills or 8 bills, and asked to determine possible amounts.

## Grades 3–5

**Six, Nine, and Twenty**

- Suppose a fast food chain sells chicken tenders only in boxes of 6, 9, and 20.

? What exact numbers of chicken tenders can you buy?
What are some exact numbers of chicken tenders you cannot buy?

***Note:*** This problem is based on one in the video *43 McNuggets* posted by Dr. James Grimes on YouTube (Grimes, 2012). It is also a good example of Mathematical Practice Standard 4, modeling with mathematics.

### *What Kind of Plan Might a Student Make?*

A student needs to realize that

- The only operation that can be used is addition (or multiplication as repeated addition).
- The only numbers that can be added in are 6, 9, and 20, but not all of them have to be used.
- The only numbers being sought are whole numbers (implicit since if you add whole numbers, you get whole numbers).

The goal is to get *all* of the numbers that are possible and some numbers that are impossible.

Some students might start with numbers that are impossible, like 1, 2, 3, 4, 5, 7, and 8. Others are more likely to start with numbers that work, since that question was asked first.

Some students will randomly choose numbers of 6s, 9s, and 20s to combine and others will be more systematic. Some will use counters and some will just add numbers. Ideally, this problem is useful to show students the power of being more systematic.

### *How Does a Teacher Push the Student Forward?*

If a student does not even know where to start, a teacher might ask:

- *Add two of the numbers that are there. What do you get? So does that number go on the list of what works or the list of what doesn't?*
- *Do you think you could get 12? How?*

- *Do you think you could get 15? How?*
- *How do you know you can get 18?*
- *If you get 18, what other numbers can you definitely get?*

If a student has found a few numbers in the 20s that work, a teacher might ask:

- *So you have 20, 21, and 26. Do any other numbers in the 20s work? Are there any you are pretty sure do not work?*

Here is an example of the work of a trio of students who did get somewhere:

50
20+9=29
29+9=38
38+9=47
56
20+6=26
26+6=32
32+6=38
38+6=44
44+6=50
6+9=15
9+20=29
6+20=26
You can get numbers 27, 50, 59.
9+9+9=27
You cant get 13, 55, 34
You can't get 28.

A teacher might respond to this solution by asking:

- *How do you know that you can't get 28 or 55?* [**Note:** This could provide an opportunity to bring in Mathematical Practice Standard 3 if students had to construct their argument and deliver it to another student to critique.]

- *Why did you write the +6s next to the 20 + 6, 26 + 6, and so forth? What does that tell you?* [**Note:** This brings in Mathematical Practice Standard 8, using and recognizing regularity in repeated reasoning.]
- *Do you think that you could get REALLY BIG numbers?*
- *How are you going to test for ALL of the numbers?*

### *How Does This Problem Build Perseverance?*

Again, this problem has many solutions, which helps build perseverance. As well, the numbers students are adding are not intimidating.

If some of the *43 McNuggets* video (Grimes, 2012) is shown in class (but not all of it), it is likely to make students want to persist to see why 43 is not possible.

Students might discover that there are many impossible numbers (e.g., 1, 2, 3, 4, 5, 7, 8, 10, 11, 13, 14, 16, 17, 19, 21, 22, 23, . . .) with 43 being the greatest. But since all of the numbers from 44 to 49 are possible:

44 as 20 + 6 + 6 + 6 + 6,
45 as 9 + 9 + 9 + 9 + 9,
46 as 20 + 20 + 6,
47 as 20 + 9 + 9 + 9,
48 as 6 + 6 + 6 + 6 + 6 + 6 + 6 + 6, and
49 as 20 + 20 + 9,

then any number greater than these is possible by simply adding 6 to each of the representations of 44–49 to get 50–55, 6 more to get 56–61, and so forth.

➢ ***Variations.*** This particular problem is actually only one in a larger set of problems called Frobenius problems, where the greatest number not possible by adding combinations of particular numbers is called for. Similar problems could be created involving scores for various sports games where only certain numbers of points can be achieved.

## Grades 6–8

**Make a Design**

- Suppose someone decides the yellow pattern block is worth $\frac{1}{2}$.

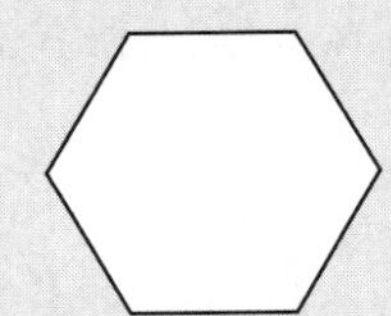

? What designs could you make that are worth $\frac{5}{3}$?

### *What Kind of Plan Might a Student Make?*

A student might realize that

- It would be useful to figure out the values of all of the other blocks in the set based on the yellow being worth $\frac{1}{2}$.
- $\frac{5}{3}$ is more than $\frac{1}{2}$, so the full design would have to be bigger than a yellow pattern block (implicit information).
- Knowing how many halves are in $\frac{5}{3}$ could be useful.

The goal is to create a design worth $\frac{5}{3}$ and justify the solution.

Some students might find the value $\frac{5}{3}$ uncomfortable. The teacher might encourage them to use a value like $\frac{3}{4}$ or $\frac{6}{4}$ first, come up with a plan, and then move toward solving the original problem. This is an example of solving a simpler problem first.

Going back to the original problem, a student might think:

*Since it takes 6 triangles to make a hexagon, each triangle is worth $\frac{1}{6}$ of $\frac{1}{2}$. That is $\frac{1}{12}$.*

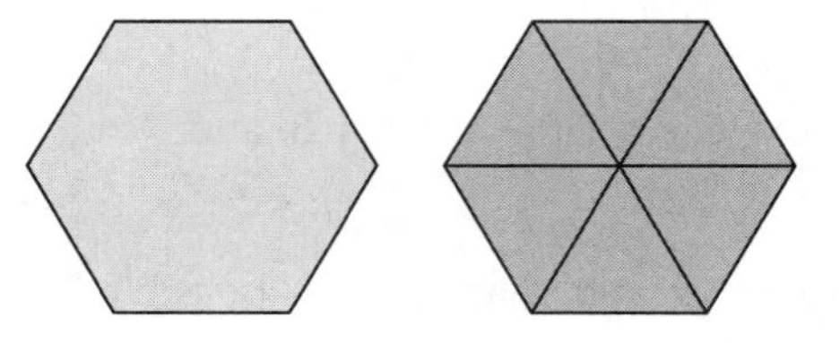

*Then, since it takes 2 triangles to make a rhombus, each rhombus must be worth $\frac{2}{12}$, and since it takes 3 triangles to make a trapezoid, each trapezoid must be worth $\frac{3}{12}$.*

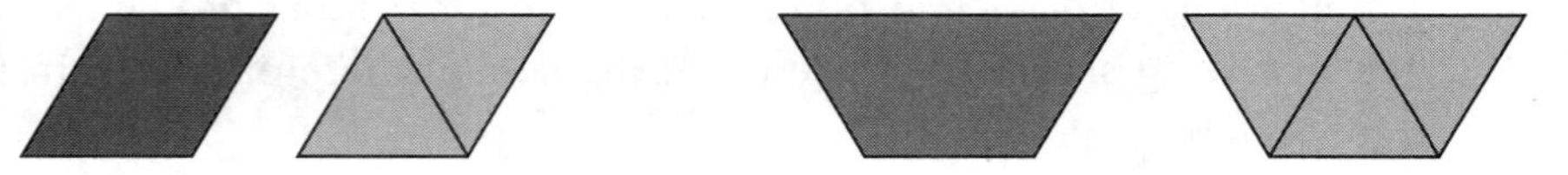

Hopefully, a student will see the value of thinking of $\frac{1}{2}$ as $\frac{6}{12}$ and $\frac{5}{3}$ as $\frac{20}{12}$ and realize that she or he needs combinations of 1, 2, 3, and 6 to make 20. That would tell what blocks to use. This is an example of Mathematical Practice Standard 2, reasoning abstractly and quantitatively. The example at the top of the next page shows how a student built on this realization:

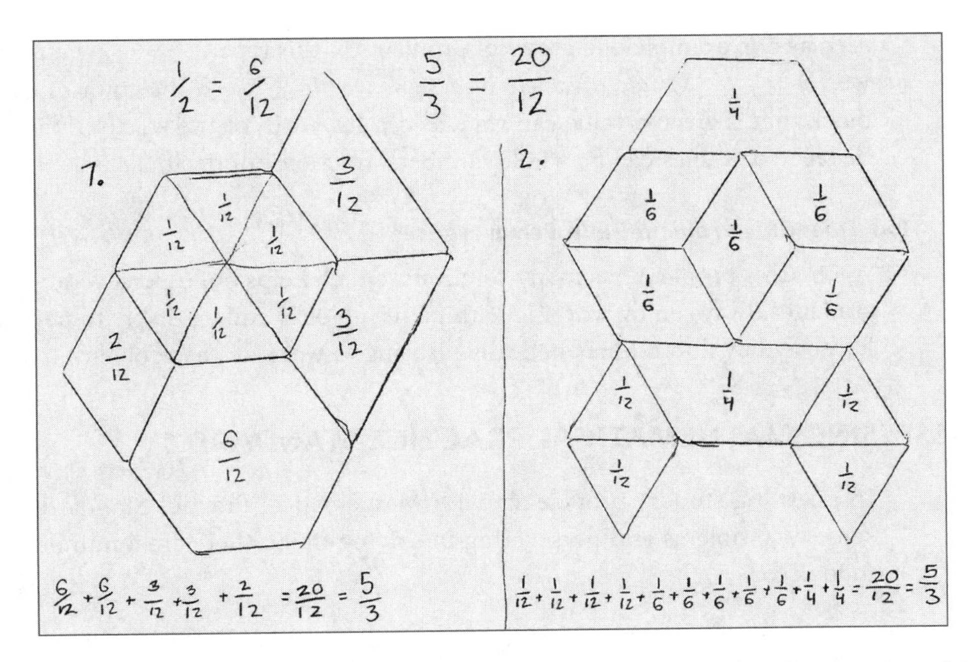


So several possible solutions are ones involving: 20 triangles, 10 rhombi, 3 hexagons and a rhombus, 6 trapezoids and a rhombus, and so forth.

### *How Does a Teacher Push the Student Forward?*

But what if a student doesn't come up with any ideas, or doesn't even know how to start? A teacher might ask:

- *Could you use 4 hexagons? Why not?*
- *Would 1 hexagon be enough?*
- *What value do you think the trapezoid would have?*

And what if a student has come up with one or two solutions, but that's all? A teacher might ask:

- *Do you think you have all the possible solutions? Why not?*
- *How could you get another solution?*
- *I notice you have two solutions. How could you exchange blocks and get more solutions?*
- *I notice you have been adding to solve the problem, which certainly has worked. I wonder how you might have used dividing to help you.* [**Note:** Figuring out that $\frac{5}{3} \div \frac{1}{2} = \frac{10}{3}$ or $3\frac{1}{3}$ would have told a student he or she needed the equivalent of $3\frac{1}{3}$ hexagons, or 3 hexagons and a rhombus.]

This problem is valuable in and of itself in clarifying the meaning of fraction division as well as in practicing fraction operations. But it is also valuable when seen as one example of a group of problems of this type: *A ______ pattern block is worth ______. Create a variety of designs worth ______.* The color of the pattern block that is given a value can vary, as can its worth or the worth of the design to be created. Values can be whole numbers, fractions, or decimals.

### *How Does This Problem Build Perseverance?*

Again, this problem has many solutions, which helps build perseverance. As well, students always enjoy working with pattern blocks and are likely to persist simply because they find it kinesthetically pleasing to work on the problem.

## ASSESSING MATHEMATICAL PRACTICE STANDARD 1

In assessing student proficiency with Mathematical Practice Standard 1, making sense of problems and persevering in solving them, there are a number of things to look for:

- Is the student recognizing what the givens are, whether implicit or explicit?
- Does the student notice if there are contradictory or missing givens?
- If the student makes assumptions, is she or he aware that those assumptions have been made?
- Does the student make a plan that makes sense given the nature of the problem?
- Does the student use a variety of strategies or does he or she always fall back on guess and test?
- Does the student show at least some independence in pursuing the problem?
- Does the student persevere for a reasonable amount of time or does she or he quit quickly?
- Does the student regularly check to see if his or her answer makes sense by either using an alternate strategy or applying logical reasoning?

## SUMMARY

In order for students to engage in Mathematical Practice Standard 1:

- A teacher must present a problem that truly is a problem but is not too far beyond students' existing knowledge.
- Often, the problem should have many solutions in order to encourage perseverance.

- A teacher must encourage students to think about what information they know, whether it is explicit information or implicit information, and what they need to figure out.
- A teacher must talk about what conjectures are and encourage students to regularly make conjectures.
- A teacher must ask scaffolding, probing, and extending questions.
- Students must be encouraged to check their answers to ensure they make sense.
- Students must have opportunities to see other solutions and learn from them.

Good problems often span many grade levels and are simply approached differently at different levels.

Good problems often bring out other Mathematical Practice Standards beyond Standard 1.

# • CHAPTER 2 •

# Reasoning Abstractly and Quantitatively

## *Mathematical Practice Standard 2*

| ***MP2. Reason abstractly and quantitatively.*** |
| --- |
| Mathematically proficient students make sense of quantities and their relationships in problem situations. They bring two complementary abilities to bear on problems involving quantitative relationships: the ability to decontextualize—to abstract a given situation and represent it symbolically and manipulate the representing symbols as if they have a life of their own, without necessarily attending to their referents—and the ability to contextualize, to pause as needed during the manipulation process in order to probe into the referents for the symbols involved. Quantitative reasoning entails habits of creating a coherent representation of the problem at hand; considering the units involved; attending to the meaning of quantities, not just how to compute them; and knowing and flexibly using different properties of operations and objects. |

**ALTHOUGH** some view mathematics as primarily about problem solving, others see reasoning as the heart of mathematics. Mathematical Practice Standard 2 focuses on reasoning, clearly an important element of mathematics. But it also focuses on the connections between mathematics and real-world problems in its emphasis on decontextualizing and then recontextualizing.

## WHEN WOULD THIS PRACTICE APPLY?

This mathematical practice standard applies in various types of situations. Sometimes students are confronted with contextual problems and need to work out a solution, often appealing to abstract mathematical strategies. But at other times, they need to resolve quantitative relationship situations that may or may not be contextual.

## CONTEXTUAL SITUATIONS

In a contextual situation, students are presented with information about a real-life situation and use mathematics to help resolve a problem. In the very earliest grades, the situations likely involve counting, addition, or subtraction. The situations become more complex and might involve other operations or algebraic expressions as students get older.

Rather than telling students how to resolve a particular type of problem, there is great benefit in letting them try to figure things out for themselves. Here I present examples of problems, showing how they would be decontextualized with mathematics and then recontextualized to explain their solutions. What decontextualizing/recontextualizing means is that a situation that may not be presented in a mathematical format is "translated" into a mathematical format consistent with the situation. Then mathematical tools are applied to lead to a solution described in terms of the original situation. In the simplest form of decontextualizing/recontextualizing, a student might be faced with a problem involving 3 bowls of 5 apples each; this is represented mathematically as $3 \times 5$. Mathematical strategies are employed to realize that $3 \times 5 = 15$. And then the 15 is described in terms of the situation, that is, there are 15 apples.

## EXAMPLES OF PROBLEMS INVOLVING CONTEXTUAL SITUATIONS THAT MIGHT BRING OUT MATHEMATICAL PRACTICE STANDARD 2

### Grades K–2

**Shopping on a Budget**

- Choose three prices, each $10 or less, for three small items you want to buy.

? Would it cost more to buy the most expensive item or both of the two cheaper items?
How much more or how much less?

One of the strengths of this open-ended problem is that it allows for multiple entry points, since students can choose whatever numbers they are comfortable with. But it also allows for a lot of computational practice, since students can be encouraged to try many different combinations of prices. As well, it helps students come to the conclusion that when you combine small numbers, the result is still small, but combinations of medium-sized numbers can get bigger than you might have first thought.

To decontextualize the problem, students might choose three values to represent the prices, add the two smaller amounts, and then either subtract the result from the greatest amount or subtract the greatest amount from the sum of the two smaller amounts. (This is where students need to recontextualize to see which way to subtract.)

For example, if the selected prices are \$3, \$4, and \$8, the student needs to add $3 + 4 = 7$ and then needs to consider $8 - 7 = 1$. Or the student might decontextualize using $8 - 3 - 4 = 1$. But if the prices are \$3, \$4, and \$6, the student needs to subtract 6 from 7 (3 + 4), not the other way around.

Clearly, Mathematical Practice Standard 1 is also applied as students figure out what to do and persist in doing it, particularly if several possibilities are requested.

➤ ***Variations.*** A teacher can increase the challenge by asking what all of the possible results are and how each can be achieved. For example, a result of \$0 could be achieved if the prices were \$2, \$4, and \$6. A result of \$1 could be achieved with prices of \$3, \$4, and \$8. A result of \$2 could be achieved with prices of \$3, \$4, and \$9, and so forth. Although most students are likely to use whole number prices, students can also choose to use dollars and cents, if they wish.

## Grades 3–5

**Is It Possible?**

? Is it possible to spend exactly \$100 and buy only items that cost either \$3 or \$6?

This problem is an interesting one because not only is there a specific solution to a specific problem, but there is a bigger generalization lurking just below the surface. Students can learn that if you add multiples of 3 (which combinations of 3 and 6 are), the result will always be a multiple of 3. This is, in fact, an application of the distributive property of multiplication over addition and an application of Mathematical Practice Standard 7, recognizing and using structure.

In solving this problem, some students might decontextualize by writing the equation $3 \times \square + 6 \times \triangle = 100$ and substitute different numbers for the unknowns in an effort to resolve the problem. Then students need to recontextualize, realizing that there cannot be a whole number answer to this problem since the value on the left is a whole number multiple of 3, but 100 is not.

Or some students might try to figure out a multiple of 3 to add to a multiple of 6 to reach a total near 100, adjusting when they don't achieve exactly 100, until they realize it won't happen.

➤ ***Variations.*** This problem can be generalized to something of the form, "Is it possible to spend exactly \$______ and buy only items that cost \$______ and \$______," using any three desired whole number values (or even decimal values later on).

## Grades 6–8

### A Great Sale

- You buy an item on sale and save $12.
- The original price was between $20 and $100.

? What could the original cost and the percent off have been?

What is good about this problem is that there are many answers, some fairly simple to allow easy entry, and some much more complicated, suited to students ready to handle them.

One student might choose a savings percentage of 50%. He or she would calculate what $12 is 50% of by using the equation $\frac{x}{2} = 12$ or perhaps $0.5x = 12$. Solving that equation yields the value $x = 24$. Then students must recontextualize, realizing that 24 represents the dollar value of the original price when the discount is 50%.

Yet another student might decide that the original price was $100 and recognize that the savings is $12; this time the unknown will represent the percent and not a dollar value. This student's equation might be $100x = 12$. The solution, $x = \frac{12}{100}$, tells that the discount is 12%. Deciding the units to use when the equation is solved is where the recontextualization occurs.

➤ ***Variations.*** This problem can be adapted by changing the amount saved and the original price range.

### Text Messages

? How many text messages do you send in a year?

This problem emphasizes proportional reasoning and is certainly of interest to students. Two particularly valuable features of this problem are that (1) the strategies used generalize to many other counting problems that require proportional reasoning and (2) students use both implicit and explicit givens to solve the problem.

Most students use information about how many text messages they send in a day and then generalize it to a year. Those students would take their number of messages per day, multiply that number by 365, and then interpret the result as the number of messages in a year.

Other students would, instead, choose to be more precise. They might divide up the year into weekdays and weekends, if they felt the messaging was very different in these two situations. They might multiply one number by 5 (for weekdays), add another number multiplied by 2 (for weekends), and then multiply that total by 52. Or other students might divide the year into school-year months and out-of-school months, again because they feel the numbers of daily messages are very different in those situations. Regardless of the approach, deciding what numbers to multiply and/or add leads to a decontextualized calculation, which is then recontextualized in its interpretation.

## NUMBER RELATIONSHIPS

Another place where this practice standard is useful is with problems involving number relationships. Although these problems do not involve real-life context, they focus on generalizations and require reasoning abstractly and quantitatively. Some examples are shown below.

## EXAMPLES OF PROBLEMS INVOLVING NUMBER RELATIONSHIPS THAT MIGHT BRING OUT MATHEMATICAL PRACTICE STANDARD 2

### Grades K–2

**Which Numbers?**

- Choose a number.
- Find the number that is 4 more than your first number. Call this number A.
- Subtract 1 from your first number.
- Find the number that is 6 more than this new number. Call this number B.

? How much greater is A than B or B than A?
Does it depend on the first number you chose?

Students can choose a comfortable number to start with. For example, if a student chooses 3, she or he might realize that A = 3 + 4. And B = 2 + 6. So 8 (number B) is 1 more than 7 (number A). But if the student starts with 6 instead, number A is 6 + 4 and number B is 5 + 6 and B is still 1 more than A. Students could use Mathematical Practice Standard 3 and explore a conjecture that B will always be 1 more than A and investigate why.

Students might use concrete objects to represent the problem. Imagine a pan balance like this:

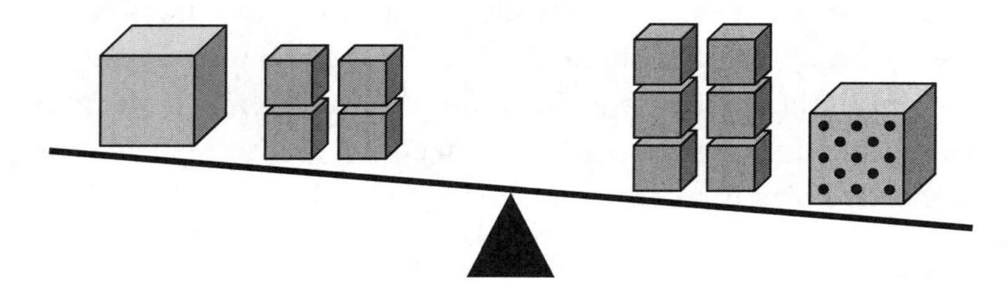

One can see that if the number of blocks in the big gray box is 1 more than the number in the spotted box, the balance could actually be thought of this way:

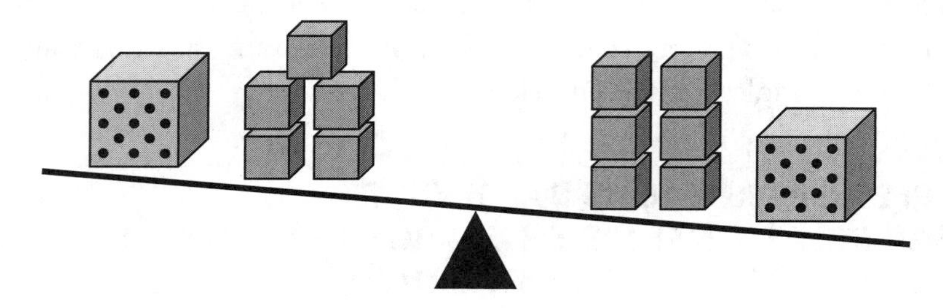

In this version it becomes clear that the right side is 1 more than the left side. The decontextualization this time is thinking about a "random" box instead of a particular number.

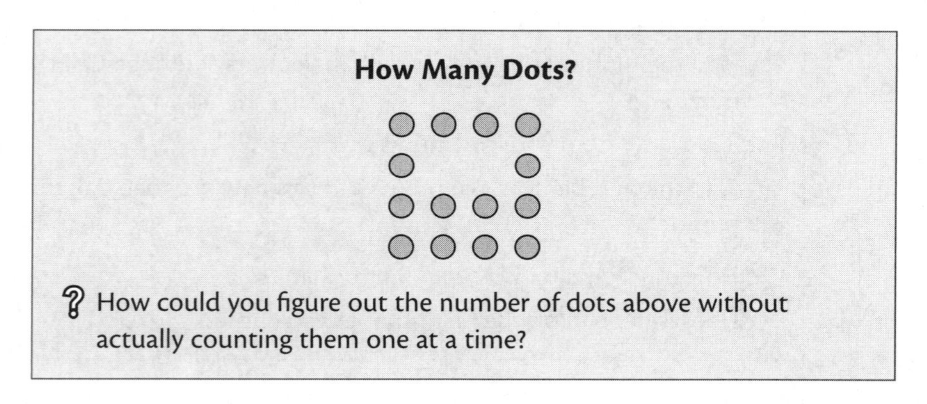


This is a specific example of a "dot problem." These types of problems are useful to call upon on a regular basis with students. They help students see numbers as composed of other numbers in different ways.

For example, in this case, the student might see 8 dots at the bottom to add to 2 on the left side, 2 on the right, and 2 more in the middle of the top. With this view, $8 + 2 + 2 + 2 = 14$ tells the total number of dots. On the other hand, the student might see 4 in the leftmost column, 3 in the second column, 3 in the third column, and 4 in the rightmost column, and that student's total would be $4 + 3 + 3 + 4 = 14$, which is, of course, the same total.

Or a student might see three rows of 4 and 2 more dots in row 2, for a total of $4 + 4 + 4 + 2 = 14$, as well.

Older students might see $4 \times 4 - 2$, which is also 14.

➢ ***Variations.*** It is possible to create many of these sorts of problems by beginning with an array of any size and strategically removing dots. For example, a teacher might start with the following:

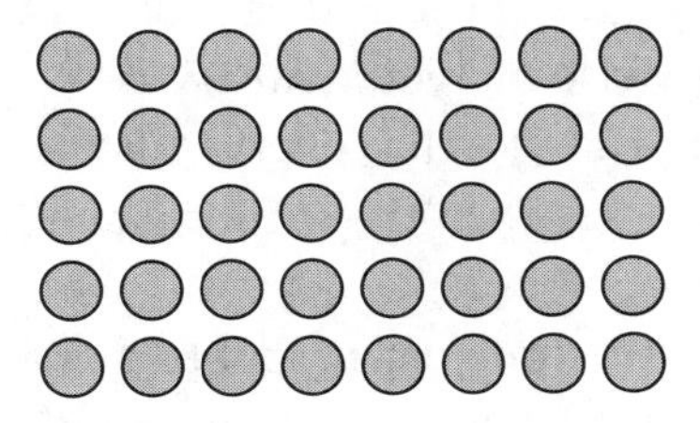

Removing dots could show any of these arrays, and more:

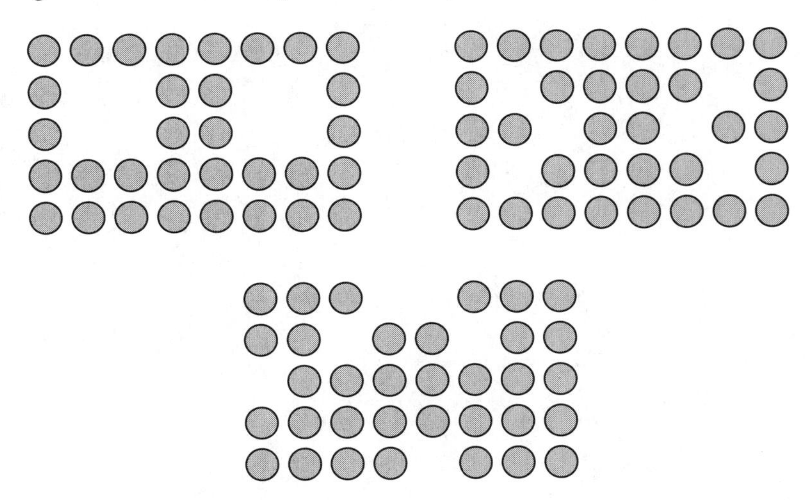

## Grades 3–5

**Two Truths and a Lie**

? Which of these is the lie? How do you know?

A. The number 68 can be represented with 32 base ten blocks.

B. The number 148 can be represented with 43 base ten blocks.

C. The number 502 can be represented with 142 base ten blocks.

This problem is a good one to give students practice with place value concepts. Initially, students might wonder what is going on. Isn't 68 represented with 6 ten-rods and 8 one-blocks since $68 = 6 \times 10 + 8$ (or $10 + 10 + 10 + 10 + 10 + 10 + 1 + 1 + 1 + 1 + 1 + 1 + 1 + 1$)? But that is only 14 blocks. But then students have to think harder: How else can you write 68 as the sum of tens and ones? Maybe $10 + 10 + 10 + 10 + 10 + 1 + 1 + 1 + 1 + 1 + 1 + 1 + 1 + 1 + 1 + 1 + 1 + 1 + 1 + 1 + 1 + 1 + 1$. That's 23 blocks. Then they think again. It could be 4 ten-blocks ($10 + 10 + 10 + 10$) and 28 one blocks ($1 + 1 + 1 + \ldots + 1 + 1$). And that is 32 blocks.

So each time, the problem is about whether the number can be written as a combination of a certain number of hundreds, tens, and ones.

As shown below, this student discovered that statement B is the lie:

B is the lie. I know because I used 4 tens witch equals 40 and then I used 28 ones. 40+28=68. 4+28=32. I also know because on, C I useded 40 ten blocks witch eqaules 400. Then I used 102 ones. 40+102=142
400+102=502.
I tried to get an anwser for B but I didn't get one.

➢ ***Variations.*** As with other number relationship problems, this problem is only one example of many, since any whole number (or even decimal) can be represented in a number of ways.

**What Could It Be?**

- You add three consecutive whole numbers (numbers in a row, such as 4, 5, and 6).

? Which of these answers are possible? Which are not? Why?

21 48 102 300 400

While students are likely to randomly select consecutive numbers to add and see what happens, hopefully they will have developed habits of mind based on Mathematical Practice Standard 1, making sense of problems and persevering in solving them, that will lead them to look for relationships in what they discover and to make conjectures.

A student might add 10 + 11 + 12 = 33, then 20 + 21 + 22 = 63, and hypothesize that the sum of three consecutive numbers has to be a number ending in 3, and thus conclude that none of the values listed are possible. This happens to be incorrect.

A teacher then needs to prompt, for example:

- *What if you tried 14 + 15 + 16? Does that end in a 3?*

The student whose work is shown below came up with a good solution, but it doesn't quite tell us how she knows that 400 is impossible just because 399 is possible.

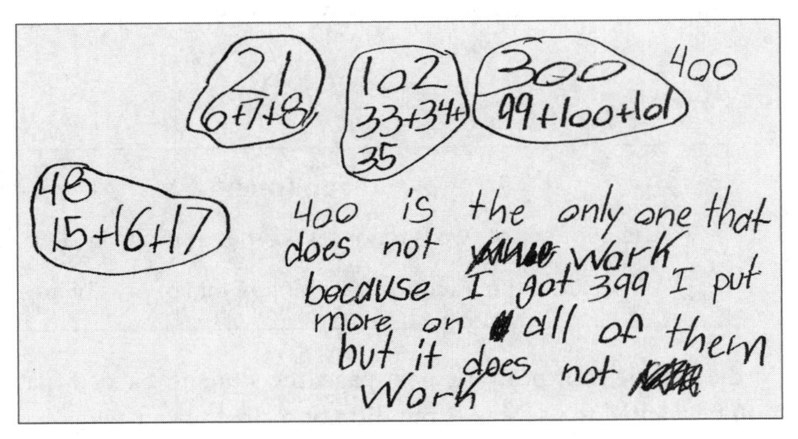


Ideally, students will start to notice that each of the numbers they get is the same as adding three of the middle number, for example, 10 + 11 + 12 = 11 + 11 + 11, or 5 + 6 + 7 = 6 + 6 + 6.

Students could explore why this is the case. (This is an example of the associative property of addition, combined with the zero property since, for example,

$10 + 11 + 12 = 11 - 1 + 11 + 11 + 1$, which is $11 + 11 + 11 + 0$, and relates to Mathematical Practice Standard 7, recognizing and using structure.)

They might even explore this type of problem visually, a different way to decontextualize. For example, a student might represent $5 + 6 + 7$ in this way:

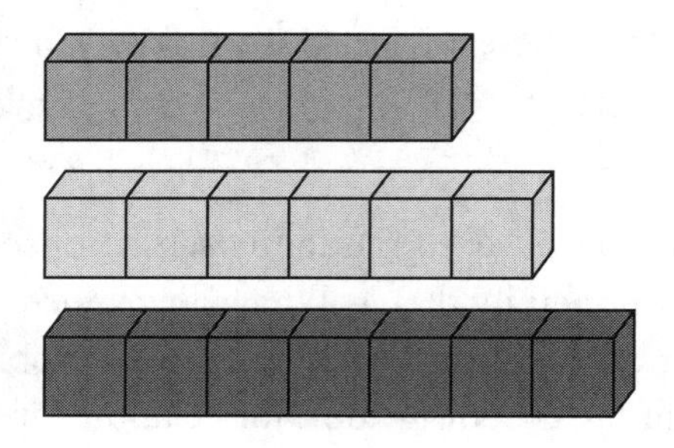

This can quickly be seen to be the same as three 6s:

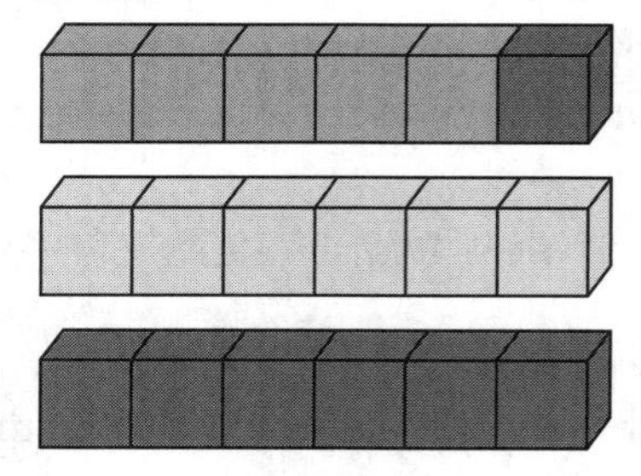

➢ ***Variations.*** Other problems could involve the sum of five, seven, or nine consecutive numbers (which are multiples of 5, 7, and 9, respectively).

## Grades 6–8

**Four Times**

- The quotient of two fractions is 4 times as much as their product.

? What could the fractions be? Think of lots of possibilities.

In the sample shown on the next page, the student realized that the second fraction he wanted to use was $\frac{1}{2}$, but did not really indicate why:

$$\frac{a}{b} \div \frac{c}{d} \overset{\times 4}{>} \frac{a}{b} \times \frac{c}{d}$$

$$\frac{a}{b} \times \frac{d}{c} \overset{\times 4}{>} \frac{a}{b} \times \frac{c}{d}$$

$$\frac{1}{2} \times \frac{2}{1} \overset{\times 4}{>} \frac{1}{2} \times \frac{1}{2}$$

$$\frac{2}{2} \overset{\times 4}{>} \frac{1}{4}$$

$$1 \overset{\times 4}{>} \frac{1}{4}$$

---

$$\frac{2}{4} \times \frac{4}{2} \overset{\times 4}{>} \frac{2}{4} \times \frac{2}{4}$$

$$\frac{8}{8} \overset{\times 4}{>} \frac{4}{18}$$

$$1 \overset{\times 4}{>} \frac{1}{4}$$

Some students will choose to use algebra to either get the answer or prove why the number to multiply or divide by is $\frac{1}{2}$. For example, the student above might have noted that $\frac{d}{c}$ must be 4 times $\frac{c}{d}$, so $d^2 = 4c^2$, and $\frac{c}{d} = \frac{1}{2}$.

Another student might approach the problem numerically rather than algebraically. He or she might think:

> *If a number is divided by $\frac{1}{2}$, the quotient is the number of halves in the number (which is 2 times the number) and the product of that number multiplied by $\frac{1}{2}$ is half of the number. That means twice a number is always 4 times half the number, so the second fraction should be $\frac{1}{2}$.*

➢ ***Variations.*** This sort of problem can be generalized to, for example, the quotient is 9 times as much or 16 times as much or perhaps $\frac{9}{4}$ as much as a product.

## ASSESSING MATHEMATICAL PRACTICE STANDARD 2

In assessing student proficiency with Mathematical Practice Standard 2, reasoning abstractly and quantitatively, there are a number of things to look for:

- When there is a contextual situation, is the student's mathematical representation of it appropriate?
- Does the student interpret the mathematical results of his or her computations in light of the contextual situation when there is a contextual situation?
- When the situation is a generalization involving number relationships, does the student use appropriate properties and meanings of operations and sufficient breadth and types of examples to draw the conclusions she or he does?

## SUMMARY

In order for students to engage in Mathematical Practice Standard 2:

- There is value in using both contextual problems and number relationship problems; both types can be symbolized with mathematics. Ideally, the problems should be generalizable.
- When students represent the mathematics of their solution, the representation can be numerical or algebraic, but it can also be visual.
- Whether students use numerical, algebraic, or visual representations, they should be queried frequently on what the symbols they use represent and what their result represents.

# • CHAPTER 3 •

# Constructing and Critiquing Arguments

## *Mathematical Practice Standard 3*

***MP3. Construct viable arguments and critique the reasoning of others.***

Mathematically proficient students understand and use stated assumptions, definitions, and previously established results in constructing arguments. They make conjectures and build a logical progression of statements to explore the truth of their conjectures. They are able to analyze situations by breaking them into cases, and can recognize and use counterexamples. They justify their conclusions, communicate them to others, and respond to the arguments of others. They reason inductively about data, making plausible arguments that take into account the context from which the data arose. Mathematically proficient students are also able to compare the effectiveness of two plausible arguments, distinguish correct logic or reasoning from that which is flawed, and—if there is a flaw in an argument—explain what it is. Elementary students can construct arguments using concrete referents such as objects, drawings, diagrams, and actions. Such arguments can make sense and be correct, even though they are not generalized or made formal until later grades. Later, students learn to determine domains to which an argument applies. Students at all grades can listen or read the arguments of others, decide whether they make sense, and ask useful questions to clarify or improve the arguments.

**THIS STANDARD**, somewhat like the previous one (Mathematical Practice Standard 2, reasoning abstractly and quantitatively), has at its heart the importance of reasoning in the world of mathematics.

## WHAT MIGHT THE ARGUMENTS BE ABOUT?

One type of argument that a student might make is one that defends a solution to a specific problem by explaining how he or she solved it. But it might be even more worthwhile if the argument represented a class of problems or a bigger idea that students are learning about. Examples of those sorts of arguments at many grade levels are included in this chapter.

## WHAT FACTORS INTO A QUALITY ARGUMENT?

There are a number of issues that should, over time, be brought to students' attention as contributing to the quality of a good argument:

- The argument is not based on just a few unrelated examples.
- The argument is not circular. In other words, a student does not say, for example, that you can add two numbers in any order since the order does not matter, thus using the result to be proved as its explanation.
- The argument is often accompanied by strong visual support.
- The argument addresses the underlying meaning of the elements of the problem, and not just the surface features. For example, a good argument about why the sum of two evens is even does not just offer a lot of examples; it is more about what the term "even" means.

The ability to create a sound mathematical argument is learned over time. Discussing the characteristics of a good argument listed above (i.e., not based on unrelated examples, not circular, etc.), as these issues arise, is critical to helping students improve their reasoning skills.

For many adults and students, arguments that use fancier mathematical terms or symbols might appear to be stronger arguments, but this is not always the case. Sometimes very strong arguments are based on a picture and a few words accompanying it.

For many adults and students, tight organization is a critical part of an argument. As students get older, this might be more of an expectation, but it is certainly not required for younger students.

## WHAT DO GOOD CRITIQUES OF ARGUMENTS SOUND LIKE?

As well as learning how to create arguments, students need opportunities to learn how to ask questions of other students' arguments. Initially, some standard "talk moves" such as those listed below might help students start their critiques. Later, they will learn to create questions more specific to a situation.

For example, some good starters for a critique might be these:

- *I agree with ______ because ______.*
- *I didn't understand why you ______.*
- *I disagree with ______ because ______.*
- *I wonder why you ______.*
- *What if you had ______?*

## EXAMPLES OF PROBLEMS THAT MIGHT BRING OUT MATHEMATICAL PRACTICE STANDARD 3

### Grades K–2

**All about Ten**

? What do you think are the three most useful things to know about the number 10?

Why are these things most important to you?

For this problem, there is no right or wrong answer; it just asks for an opinion. But the answer is still something that could be viewed as an "argument" that could be offered and critiqued. One student might argue that the three most important things to know about 10 are that

- 10 is the last number little kids learn to count to when they start counting.
- 10 comes after 9.
- 10 is the first two-digit number.

This student might suggest that these are important features because the first one shows that a lot of people see 10 as an important number or they would not have ended the counting at 10. And the other two are important because they give a good idea of where 10 is in the list of numbers and so give a feel for how big it is.

But other students should be encouraged to question this reasoning:

- *How does knowing that 10 comes after 9 or that it is the first two-digit number give you a sense of its size?*
- *Why does it matter that it's the biggest number that little kids can say?*

➢ ***Variations.*** This type of problem can be generalized to others where students are asked to explain why they think a particular property, number, or shape is important.

**Moving on a Number Line**

- Leah says that 3 and 7 are 5 apart since if you are on a number line and you start on 3, you touch 3, 4, 5, 6, and 7; that is 5 numbers.
- Emma says that 3 and 7 are 4 apart since if you start at 3 on a number line and jump 4 spaces, you get to 7.

? Do you agree with Leah or with Emma?

Note that this situation addresses a common misconception among students (Leah's suggestion); that fact alone makes this task particularly worthwhile to discuss. Ideally, both Leah and Emma would "act out" their explanations. Then other students should be encouraged to ask questions to critique their arguments. For example, other students might ask:

- *Aren't they 3 apart since 4, 5, and 6 are between them?*
- *What do you mean when you say "apart"?*
- *But if you are already on 3, why do you count the 3, since you haven't really moved?*
- *Can both answers be right?*

**Right or Wrong?**

- Suppose 4 + 9 = □ + 7.
- Connor said that the answer is 13 since 4 + 9 = 13.

? What's wrong with his argument, or is he right?

Connor's approach is, in fact, a common error that students make. Students need to think about what it actually means to use an equal sign. Does it mean that the answer follows, which is what Connor thought, or does it mean that both sides of the equation represent the same value? Those are two very different things.

Discussion of this argument might be managed by allowing students to vote on their positions on small whiteboards that they hold up; this ensures that each student makes a decision independently. Then one student on either side of the argument could be asked to make a short presentation and try to win others over to his or her side. Ultimately, it is very important that students realize that Connor is incorrect, as the student whose work is shown below demonstrates:

Connor is wrong because 4+9 is not balanced with 13+7. 13+7 is 20 and 4+9 is 13. If 4+9 and 13+7 was on a scale, it would not be even because 20 is bigger than 13 by 7.

➢ ***Variations.*** Similar problems might involve equations where the unknown is on the left, for example, 32 – □ = 52 – 30.

## Grades 3–5

Another debatable question is shown below. The debate here most likely lies in the use of the term "more likely." The issue is what students have to do to be convincing. Do they really have to try every pair of possible numbers or just every pair up to 100? What else could they do?

**So Many Evens**

- Liz says that when you multiply two numbers, the answer is more likely to be even than odd.

? Do you agree or not? Why?

This conjecture is worth investigating for a number of reasons. First, it integrates a number of different concepts, relating probability ideas to multiplication concepts. Second, it is foundational for other conjectures about factors and multiples that students might explore later, such as these:

- *Are there more multiples of 3 that are multiples of 2 or multiples of 5?*
- *When you multiply a multiple of 3 by a multiple of 5, what can you be sure is true about your result?*

In preparing an argument for the original problem, some students might simply use a few examples from which to draw a conclusion, whereas others might undertake a deeper analysis.

One student suggested the following:

I don't agree because there is just as many odd numbers than even so it would be 50% of getting a odd number

This argument is incorrect, but many students might not even realize that, since it sounds reasonable.

A correct but not very convincing argument might be:

*I tried 3 × 5, 4 × 6, 5 × 9, 4 × 10, and 6 × 8. Mostly I got even numbers, so the answer is probably going to be even.*

Other students might be asked to critique this argument by asking questions or making statements like these:

- *But you just tried five pairs. What if you had picked different numbers?*
- *I wonder why you picked the numbers you did. Do you think that made a difference?*
- *What if you had used bigger numbers?*

A more convincing argument might be one like this:

*I looked at the multiplication table, and every second row was all even numbers and the other rows were half even. That means most of the products are even, so that makes it more likely that a product is even than odd.*

But other students should still critique this argument by asking questions such as these:

- *How do you know it was most?*
- *What if the numbers were bigger?*

Perhaps even more convincing might be the following argument, where a student argues why three-fourths of the answers are even and one-fourth are odd:

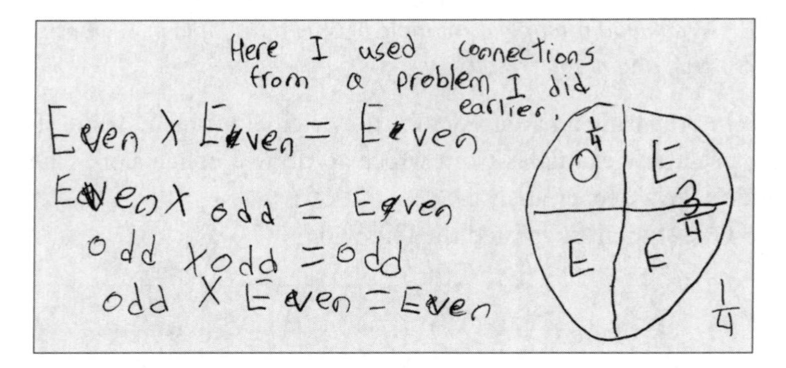


Students will eventually see that arguments based on sound deductive thinking are usually more satisfying than those based on induction.

**Which Is More?**

- Zayden says: 0.16 is more than 0.4 since 16 > 4.

? How would you reply to Zayden?

This argument deals, again, with a common misconception held by many students. Students tend to overgeneralize what they know about whole numbers and apply it, when it is inappropriate, in a decimal situation.

Notice that this time the strategy is to present an argument and ask students to critique it. We want students to realize that the number 0.16 is, in fact, less than 0.4.

Here are some possible arguments, not all as convincing as others:

- ***Argument A:*** 0.16 has only 1 tenth but 0.4 has 4 tenths, so 0.4 is more.
- ***Argument B:*** I drew pictures of both. You can see that 0.4 is more.

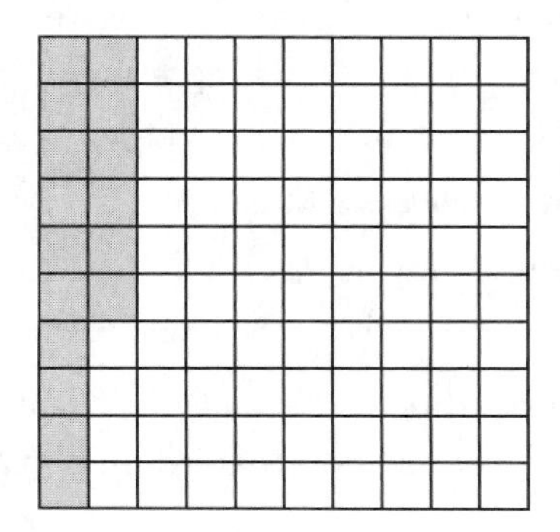
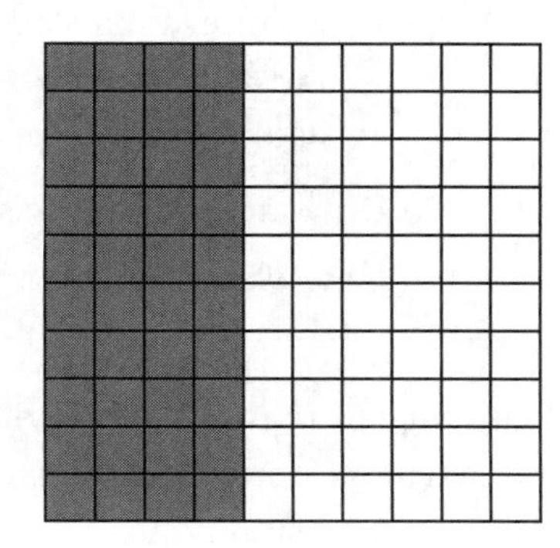

- ***Argument C:*** I know that 0.4 is 4 tenths, but $\frac{4}{10} = \frac{40}{100}$.

$$0.16 = \frac{16}{100}$$

$$\frac{40}{100} > \frac{16}{100}$$

- ***Argument D:*** 0.16 is a way to compare 16 to 100.
  16 is a really small part of 100.
  But 0.4 is a way to compare 4 to 10 and 4 is almost half of 10.
  So 0.4 must be more.
- ***Argument E:*** If you draw a number line, 0.4 is to the right of 0.16, so it is more.

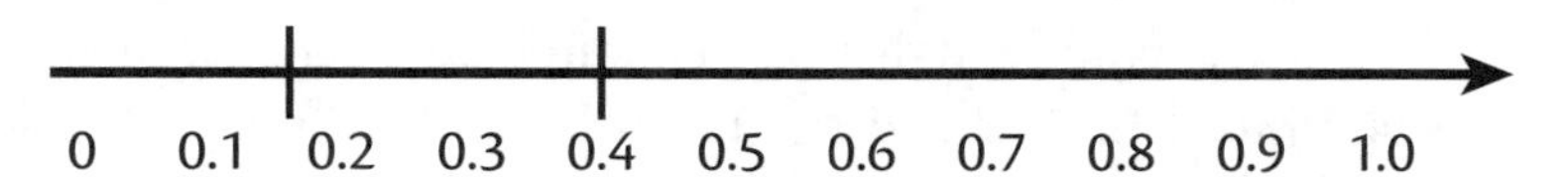


Which argument do you find most convincing?

Many students, and even adults, are attracted to Argument A because it is based on place value rules, but others might find Argument B or D more convincing.

In looking at Argument E, a student might query:

- *How did you know where to put 0.16? There was no number for it on the number line.*

The strategy of comparing arguments is valuable in many situations and at many grade levels.

**So Simple**

- Keira made this conjecture:

  To get a fraction between two fractions, just use a number between the numerators for the numerator and a number between the denominators for the denominator.

  For example, between $\frac{2}{3}$ and $\frac{8}{5}$ is $\frac{5}{4}$, since 5 is between 2 and 8 and 4 is between 3 and 5. $\frac{5}{4}$ is just a little more than 1, but $\frac{8}{5}$ is a lot more than 1 and $\frac{2}{3}$ is less than 1, so $\frac{5}{4}$ is in the middle.

? Do you agree that this strategy always works?
If so, why would that be? If not, why not?

This conjecture might be of interest to students because they really might wonder if it is true or not. It also suggests a generalization that might be valuable in future mathematical situations.

How might students argue this conjecture? Consider these possibilities:

- ***Argument A:*** I tried three pairs of fractions and it worked each time, so I believe it.
- ***Argument B:*** I tried it with two pairs of fractions less than 1, two pairs of fractions greater than 1, and two pairs where one was less than 1 and one was greater than 1. It worked each time, so I believe it.
- ***Argument C:*** First I tried fractions with common denominators, like $\frac{4}{10}$ and $\frac{6}{10}$, and it definitely worked for them since the middle of the bottom numbers was the same denominator and the middle of the top numbers was greater than the lower one and less than the higher one. Then I tried some fractions with different denominators and it seemed to work.
- ***Argument D:*** I don't think it's true. I used $\frac{8}{15}$ and $\frac{2}{3}$. I know that $\frac{8}{15}$ is less since $\frac{2}{3} = \frac{10}{15}$. For the number between 8 and 2, I used 7. For the number between 15 and 3, I used 8. So my new fraction was $\frac{7}{8}$. But $\frac{7}{8}$ is bigger than both $\frac{8}{15}$ and $\frac{2}{3}$, so you don't get an in-between fraction.

Most students (and most adults) might have been persuaded by Argument B or maybe A or C—until they heard Argument D. This shows how important it is for many students to put forth arguments and demonstrates the importance of counterexamples.

## Grades 6–8

**Big Percents**

- A store employee noticed that an item's price had been reduced by 30% and realized it was a mistake.
- So she added 30% back to the reduced price.
- Avery said the price is the same as it used to be, but Zahra disagreed.

? With whom do you agree? Why?

While one student might agree with Avery, assuming that the price must be the same since the employee undid what had been done in the first place, another might realize that Zahra is right. In the first case, the 30% was applied to a greater number than in the second case, so, in fact, the second change was not a full "undoing."

For example, if the original price had been $100, 30% of $100, or $30, would have been removed from the price. That would have made the reduced price $70. But 30% of $70 is only $21, so adding back 30% of the reduced price gives a final price of only $91 ($70 + $21), not $100.

**In Between**

- You multiply two fractions: $\frac{a}{b}$ and $\frac{c}{d}$.
- The result is a LOT MORE than $\frac{a}{b}$, but a BIT LESS than $\frac{c}{d}$.

? What could the fractions be?
Why did those fractions work?

On the surface, this problem might seem very narrow, but it actually addresses a very big idea—that when you multiply by a very small fraction, you reduce the other number's size significantly; when you multiply by a number near 1, you don't change the other number's size significantly; when you multiply by a fraction much greater than 1, you increase the other number's size significantly.

To begin the problem, a student might simply choose two fractions, for example, $\frac{2}{3}$ and $\frac{3}{4}$. She or he will soon discover the need for another path, since the desired result is not achieved. A teacher might have to prompt, for example:

- *What fractions did you try?*
- *What about some bigger fractions?*
- *What about fractions near 0?*

The student whose work is shown below discovered that, for example, $\frac{9}{10} \times \frac{10}{3}$ will work. She found this after trying a number of other possibilities, although she does not indicate how she chose the numbers.

work work

$a/b \times c/d = e/f$

way more

$e/f > a/b$

fraction?

Bit less

$e/f < c/d$

A=
B=
C=
d=
e=
f=

Answer

$\frac{10}{3} \times \frac{9}{10} = \frac{90}{30}$

$\frac{9}{10} = \frac{a}{b} = \frac{9}{10}$

$\frac{10}{3} = \frac{c}{10} = 3\frac{1}{3}$

$\frac{90}{30} = 3 \quad 3\frac{1}{3} > 3 > \frac{9}{10}$

$\frac{1}{2} \times \frac{7}{8} = \frac{28}{16} \quad 1\frac{3}{4}$ X

$\frac{1}{8} \times \frac{2}{4} = \frac{2}{32} = \frac{1}{16}$ X

$\frac{1}{10} \times \frac{5}{10} = \frac{5}{100}$ X

$\frac{1}{2} \times \frac{1}{3} = \frac{1}{6}$ X

$\frac{1}{6} \times \frac{1}{2} = \frac{1}{12}$ X

$\frac{2}{6} \times \frac{2}{3} = \frac{4}{18} \quad \frac{4}{80} < 1/3 < 2/3$

$\frac{4}{6} \times \frac{2}{3} = \frac{8}{18} \quad \frac{1}{2}$ X

$\frac{3}{6} \times \frac{2}{3} = \frac{6}{18} \quad \frac{1}{3}$ X

$\frac{7}{8} \times \frac{1}{6} = \frac{7}{48} \quad \frac{7}{8} > \frac{1}{6} > \frac{7}{48}$

$\frac{7}{8} \times \frac{2}{6} = \frac{14}{48} \quad 42 \quad 16$

$\frac{7}{8} \times \frac{2}{5} = \frac{14}{40} \quad 16$

$\frac{7}{8} \times \frac{2}{9} = \frac{14}{72} \quad \frac{16}{72}$

Another student, too, does not tell us where the numbers came from, but does a bit better job of verifying that she is correct:

The answer: I think the fractions could be....

$a/b = 9/10 \quad c/d = 5/1$

I think this because...

$\frac{9}{10} \times \frac{5}{1} = \frac{45}{10}$ meaning product of 9/10 and 5/1 is 45/10. $\frac{45}{10} - \frac{9}{10} = \frac{36}{10}$ which shows that 45/10 is a lot more then 9/10. Also $\frac{5}{1} - \frac{45}{10} = \frac{50}{10} - \frac{45}{10} = \frac{5}{10}$ meaning $\frac{45}{10}$ is a bit less then $\frac{5}{1}$. This is why I think I am right.

The important mathematics involves figuring out *why* these answers work. Ideally, a student will realize that they work because the first fraction is less than 1, but close to it, and the second is considerably more than 1.

➢ ***Variations.*** Alternative problems that are similar can easily be created. Two examples are these:

- You multiply two fractions and the result is a lot less than one of them and slightly more than the other.
- You divide two fractions and the result is slightly more than if you multiply them.

**Solving Equations**

- Lara says that the solution to

$$\frac{3}{5}x - 2 = \frac{2}{7}$$

is twice as much as the solution to

$$\frac{3}{10}x - 1 = \frac{1}{7}$$

since all the numbers in the first equation are double all the numbers in the second one.

? Do you agree or disagree? Why?

This question is valuable because it underlies all work with solving equations algebraically. Students need to realize that the answer does not change if all values in an equation are divided by or multiplied by the same value. But it is not surprising that a student would think that halving all the elements of the equation will lead to an answer that is half the original answer. Discussing such misconceptions will help to eliminate them.

Some students will pursue an argument based on actually solving both equations and noting that the solutions are the same. A stronger argument would suggest that if two amounts balance, then their doubles also balance. Students at all levels need to realize that the strongest arguments generally go back to what the mathematics is representing and not just what the solution to a particular problem happens to be.

## ASSESSING MATHEMATICAL PRACTICE STANDARD 3

In assessing student proficiency with Mathematical Practice Standard 3, constructing and critiquing arguments, there are a number of things to look for:

- Is the student aware of and careful about the assumptions made?
- Are terms well defined?
- Does the student feel comfortable making conjectures?
- Does the student recognize that a single counterexample can disprove a generalization but even 10 examples do not prove a generalization?
- Does the student use both inductive and deductive thinking?
- Does the student notice flaws in arguments?

## SUMMARY

In order for students to engage in Mathematical Practice Standard 3:

- Teachers should encourage the regular creation of conjectures that can be tested.
- Teachers should regularly require explanation of thinking.
- Students must learn how to query other students' work and their own work.
- Teachers could create situations where two different students, whether real or mythical, have taken different viewpoints on a situation and students in the class get to choose sides.
- Students should learn to question assumptions.
- Teachers might present false claims, most likely overgeneralizations of previously learned concepts, and have students respond.
- Teachers might regularly pose questions like this: *Will it still be true if . . . ?*

# • CHAPTER 4 •

# Modeling with Mathematics

## *Mathematical Practice Standard 4*

***MP4. Model with mathematics.***

Mathematically proficient students can apply the mathematics they know to solve problems arising in everyday life, society, and the workplace. In early grades, this might be as simple as writing an addition equation to describe a situation. In middle grades, a student might apply proportional reasoning to plan a school event or analyze a problem in the community. By high school, a student might use geometry to solve a design problem or use a function to describe how one quantity of interest depends on another. Mathematically proficient students who can apply what they know are comfortable making assumptions and approximations to simplify a complicated situation, realizing that these may need revision later. They are able to identify important quantities in a practical situation and map their relationships using such tools as diagrams, two-way tables, graphs, flowcharts, and formulas. They can analyze those relationships mathematically to draw conclusions. They routinely interpret their mathematical results in the context of the situation and reflect on whether the results make sense, possibly improving the model if it has not served its purpose.

**THIS STANDARD**, even more than Mathematical Practice Standard 2, reasoning abstractly and quantitatively, where relating mathematics to the real world certainly does come up, links the practice of mathematics to its applications. Students explore the utility of mathematics as a tool for solving real-life problems. A mathematical model might be more exact—for example, an addition equation of the form $3 + 2 = \square$ to show the result of adding 3 of something to 2 of something—or might be an approximation—for example, where the formula for the perimeter of a rectangle is used for a shape close to, but not exactly, a rectangle.

## ASSUMPTIONS

One of the big issues associated with this standard is the fact that when we represent real-world situations with mathematics, we are often making assumptions.

A young student might be asked to determine the number of apples he would have if he had 2 apples and got 3 more. It seems relatively straightforward to model this as $2 + 3$, but, in fact, a student has to make the assumption that the size of the apple is irrelevant when deciding to do this. As adults, we might think that this is not really an assumption since we asked how many, but this probably *is* an issue young students have to deal with.

Older students might be asked to model problems with mathematics that require more sophisticated assumptions. For example, if the teacher suggests that one child has read $3\frac{1}{2}$ times as many pages as another and asks for possible values for each child, the problem solver must decide whether the answers can be only whole numbers or can also be rational numbers; often this is not stated. Or, if students are asked to estimate the number of children in the school, they have to decide if their own classroom is representative and can be used as a factor to multiply by the number of classes, or not.

There are actually many real-world problems based on making assumptions. For example, to estimate the weight of 1000 apples, students must make estimates about the size of a typical apple. Students who are very "precise" might be uncomfortable with this sort of problem because they might realize that the variability in weights of actual apples will guarantee that $1000 \times$ the size of a typical apple will not yield a completely accurate answer. Teaching them to deal with the appropriateness of making assumptions takes time and effort.

## REASONABLENESS OF ANSWER

With Mathematical Practice Standard 2, reasoning abstractly and quantitatively, in the background, we realize that the result of solving a problem based on modeling a situation mathematically must be reconsidered in light of the context to see if it makes sense. It might mean altering the answer in a particular way, for example, changing a fraction to a whole number when dealing with a remainder situation in a context where only whole number answers make sense, but it might also mean revisiting assumptions that were made and re-evaluating those assumptions.

# EXAMPLES OF PROBLEMS THAT MIGHT BRING OUT MATHEMATICAL PRACTICE STANDARD 4

## Grades K–2

**Two Groups**

- The 18 children in Sarah's class were standing outside in two separate groups.

? If the groups were close in size, how many might have been in each group?

Young children might model this situation with an addition sentence—□ + △ = 18—and look for numbers to replace the two unknowns. They might initially begin with 9 + 9 = 18, and then would have to decide whether "close in size" allows for exactly the same size or not. Another consideration might include, for example, whether two numbers would be considered close or not by most people.

Whichever decision they make, students might then try to determine other solutions that they think meet the criteria of the problem, for example, 10 and 8 or possibly 11 and 7 (but perhaps they think 11 and 7 are not close enough).

➢ ***Variations.*** A problem like this can be generalized to situations where different totals are decomposed into subgroups with given relationships.

**Standing in Line**

- There are 23 students standing in line.
- Jack is in the middle.

? How many people are ahead of him?

Some students might use a number line or other tool to solve this problem. (Making such a decision relates to Mathematical Practice Standard 5, using tools strategically.) One student might choose a number line and then put counters at 1 and 23, representing the beginning and the end of Jack's line:

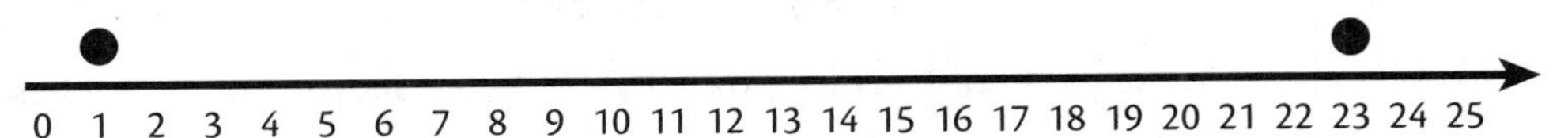


Notice that an assumption was made to put the left counter at 1 and not 0, since the first person is called 1, but this positioning had to be considered.

The student moves the counter in from the left at the same rate he or she moves the counter in from the right. Two of these steps are shown below:

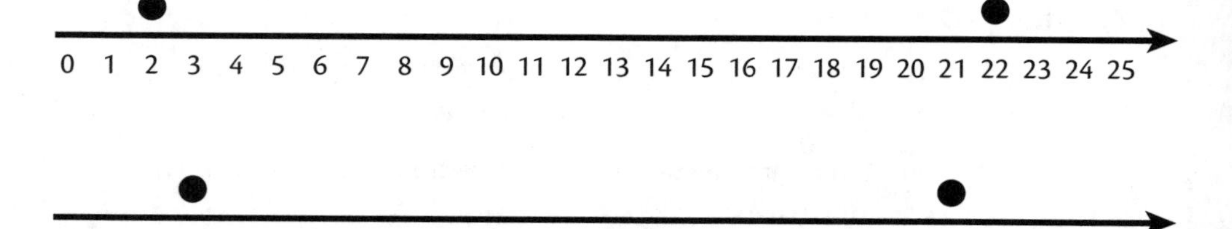


Eventually the counters meet at 12, so Jack must be at position 12, with 11 people ahead of him and 11 people behind him.

A different student might have selected a different model, for example, thinking of 23 as almost 2 equal groups with 1 extra, that is, $23 = 2 \times 11 + 1$. The two equal piles would have represented the number of people ahead of Jack and the number behind Jack.

**Paying for It**

? How can you pay for an item that costs 75¢ with exactly 6 coins?

The student has to assume that any combination of six coins is allowed, not necessarily six of the same kind of coin. Essentially, he or she has to realize that the task is to figure out how to add six numbers, each of which is either 1, 5, 10, or 25 (in the United States) or 5, 10, or 25 (in Canada), to get a total of 75.

A student might start with $75 = 25 + 25 + 25$, and realize that only three numbers have been added, not six. She or he might eventually try $10 + 10 + 10 + 10 + 10 + 25$ and determine that five dimes and a quarter is a possible solution.

## Grades 3–5

**Sharing Sandwiches**

- There are 6 students who have to share only 5 identical sandwiches.

? How much of a sandwich should each child get?

In solving this problem, students have to make assumptions. For example:

- Is it okay for a child to get a lot of little pieces or does she or he have to get just one piece, or at least not any "slivers"?
- Were the sandwiches already cut in half? Does that matter?
- Does every student need to get the same number of small pieces?

The problem does not specifically state these things, so, as adults, we can assume that the problem solver has the freedom to make all of these decisions. However, students who are younger may need permission or may be encouraged to specifically address those issues in whatever way they think makes sense, while being prepared to justify their assumptions. (Mathematical Practice Standard 3, constructing and critiquing arguments, also applies here.) This is an instance where a misunderstanding of the overall intent of standards to push students to rely solely on their own thinking might lead a teacher to say, "You need to figure that out for yourselves." In fact, however, letting students know that they have permission to make these decisions based on what makes sense for this problem can liberate them to solve it.

While one student might be perfectly comfortable dividing each of the 5 sandwiches into 6 equal pieces and giving each child a piece of each sandwich (which is mathematically $5 \times \frac{1}{6} = 5 \div 6$), other students might feel that real children would not like so many little pieces and might cut the sandwiches differently. For example, the picture below models that $5 \div 6 = \frac{5}{6} + \frac{5}{6} + \frac{5}{6} + \frac{5}{6} + \frac{5}{6} + 5 \times \frac{1}{6}$. This way, only one person gets little pieces.

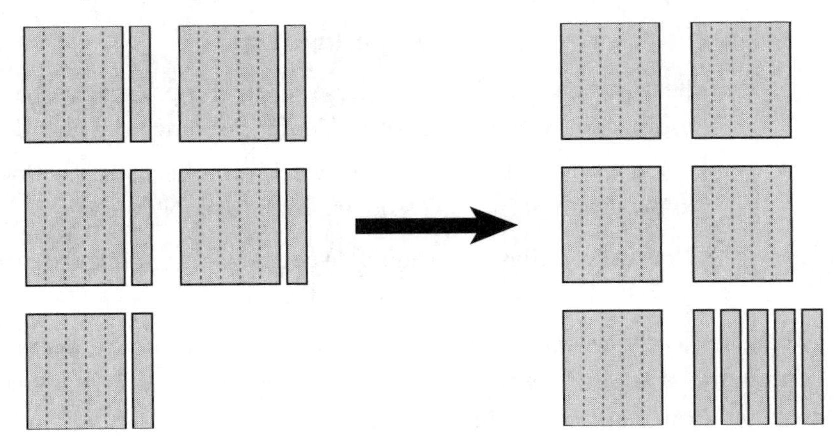

Another student might decide it is fairer and more attractive for each person to get a large piece and some smaller pieces. So each student would receive a half sandwich and small pieces from equal division of the remaining 4 half sandwiches. This problem offers students another example of how multiple answers can be correct.

**A Perfect Garden**

- You have 22 feet of fencing and want to make a garden with an area of at least 20 square feet.
- You want the garden to be close to rectangular.
- You are going to fence it all around.

? What could its length and width be so you would have enough area and enough fencing?

Some more proficient students might model this mathematically if they realize that they need dimensions $l$ and $w$ such that the sum of $l$ and $w$ is 11 or less and the product is 20 or more.

Other students might experiment with numbers that would meet these criteria. They would discover, for example, that 10 and 5 do not work since there is not enough fencing, but 6 and 5 do work. So the garden could be 6' × 5'. Or they might discover that 6 and 4 also work, although 6 and 3 do not.

Younger students might draw a rectangle and write numbers next to two sides of the rectangle and multiply them. Even when students come upon one answer that solves the problem, they might try other combinations to see what would give them the largest garden possible. Teachers may want to prompt for this additional investigation.

**Ordering Pizza**

- Because some students are away on a field trip, a teacher has to estimate how many pizzas to order for hot lunch the next day.
- She does not have access to last week's numbers, but she does know that there are 512 students in the school this week.

? How do you think she should decide how many pizzas to order?

There is clearly a need for assumptions in this problem. Students have to make assumptions about how many students will eat pizza, how many slices each will eat, and how many slices are in a pizza.

One solution might go something like this:

*In our class, only 4 people never eat pizza, and we have 21 people in our class, so that's about $\frac{1}{5}$ who don't eat pizza. I will guess it's the same for the whole school, so I will only order pizza for about 400 students. If each student ate 2 slices, that would be 800 slices. I am pretty sure that a pizza comes in 8 slices, so I would order 100 pizzas.*

This provides a great opportunity to bring in Mathematical Practice Standard 3, constructing and critiquing arguments, since other students could query whether this student's class was typical, whether $\frac{4}{5}$ of 512 is 400 or a little more and whether the little more matters, whether pizzas really do come in 8 slices, and whether or not there should be a "cushion."

## Grades 6–8

**Good Sales**

- You buy a jacket at 40% off.
- You buy shoes at 20% off.
- You pay the same amount for both items.

? What do you know about the relationship between the two pre-sale prices?

This problem is an excellent one for a lot of reasons. First of all, it can be solved numerically, algebraically, or visually, giving students real choice in accessing the problem (and bringing out Mathematical Practice Standard 5, using tools strategically). Of equal importance is that the problem emphasizes the big idea that percent relationships are multiplicative, not additive, and therefore the relationship of the values must be a multiplicative one. As a side note, the problem also emphasizes that knowing a sale percentage implicitly, but not explicitly, gives the cost percentage.

The quickest model is probably an algebraic one. If the original jacket price is represented as $x$ and the original shoe price as $y$, then $0.6x = 0.8y$. As a consequence, $6x = 8y$ and $y = \frac{3}{4}x$. The original shoe price had to be $\frac{3}{4}$ of the original jacket price.

But visual models are also possible. One visual model is shown below, where the jacket price percentages are shown on one number line and the shoe price percentages on another:

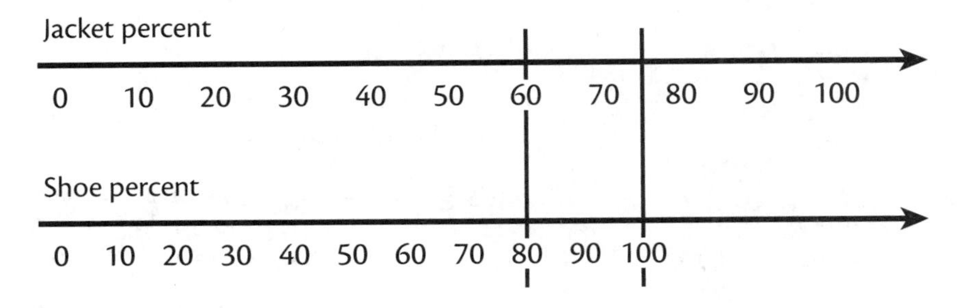

In order to have the sale prices match, percentages on the two number lines are spaced so that 60% of the original jacket price lines up with 80% of the original shoe price. This then shows that 100% of the original shoe price matches 75% of the original jacket price.

Or a student could solve the problem numerically by trying different possible jacket prices, determining the matching shoe price, and looking for relationships. It is the looking for relationships that is the tough part, since students must consider lots of possible relationships.

For example, if an original \$100 jacket price is assumed, then its sale price is \$60, which is also the sale price of the shoes. To get the original price of the shoes, the student has to determine that \$60 is 80%, or $\frac{4}{5}$, of \$75. So the price of \$75 for the shoes matches the price of \$100 for the jacket. Many students at this point would simply conclude that the shoes cost \$25 less than the jacket. They would have to be encouraged to explore, for example, prices of \$100 and \$125 to see if they work, which they don't. The students need to look for alternative relationships (not just difference) between 75 and 100 that will make the problem work. The relationship that is critical, in the end, is that 75 is $\frac{3}{4}$ of 100.

The student whose work is shown below solved a proportion to get one solution, but did not generalize. This shows the need for really focusing in on Mathematical Practice Standard 8, using repeated reasoning.

Jacket = 40% off
Shoes = 20% off
both with the prices of the sales now equal x amount

The Jacket would equal more money for it's orgional price because it needs a higher percentage of a sale off than the shoes.

The Jacket and the shoes have a 20% difference

lets say with the discount the both equal \$50

than $\frac{\$50}{x} = \frac{60\%}{100\%}$   $\frac{\$50}{x} = \frac{80\%}{100}$

\$83.3   \$62.5

Jacket
discount price - \$50
orgional price \$83.3

Shoes
discount price - \$50
orgional price = \$62.5

∴ The Jacket does equal more for it's orgional price.

\$83.3 - \$62.5
= \$20.8

Also with \$50 as discount price for both items, their orgional price has a \$20.8 difference

The student whose work is shown below came to a very interesting generalization, although it was based on relatively few examples:

The Jacket is more exspensive than the shoes because you got more money off for the jacket but still ended up paying the same amount.
40% of 10 = 4 dollars off
20% of 10 = 2 dollars off
The numbers have to be even. They have to have a difference of $2^x$

\$16 Jacket 9.60 ($2^2$)
\$12 Shoes 9.60

\$8 Jacket 4.8 ($2^1$)
\$6 Shoes 4.8

\$32 Jacket 19.20 ($2^3$)
\$24 Shoes 19.20

\$4 Jacket 2.4 ($2^0$)
\$3 Shoes 2.4

The shoes number has to be a multiple of 3. The Jacket number has to be a multiple of 4.

**Better Buy**

- One gym charges \$50 to join and \$4 for each hour the member uses the gym.
- Another gym does not require membership and just charges \$8 per hour.

? Which would be a better plan and when?

The situation described in this problem could certainly be modeled mathematically. The first membership plan might be graphed using the equation $c = 50 + 4h$. The second might be graphed using the equation $c = 8h$. The student could see when each graph is higher than the other.

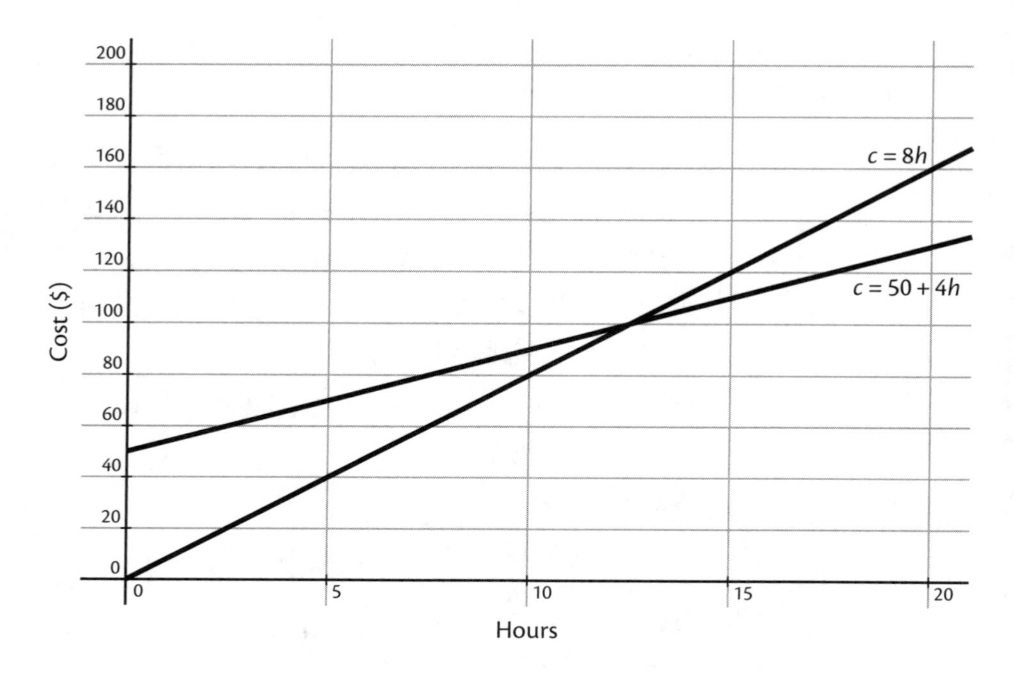


It becomes clear that the first plan is pricier unless you spend about 12 or 13 hours or more at the gym, in which case the second plan is the better buy.

➤ ***Variations.*** This is an example of the *many* problems in which students could use a linear system of equations to investigate two linear relationships between two variables.

**Pasta All Around**

- The owner of a pizza parlor wants to provide a bowl of spaghetti to every single person in Chicago, Illinois.

? How many packages of spaghetti noodles would he have to use? How much would it cost?

To model this problem mathematically requires a lot of assumption making and information gathering. What size package? How much is in a bowl? Exactly how many people live in Chicago? Should people who are in hospitals and who can't eat be counted? Should babies who don't eat solid food yet be counted? Should people who don't like spaghetti be counted? And if not, what numbers of people do I leave out? Is it cheap pasta or expensive pasta?

This sort of problem is engaging for most students, but does involve a lot of mathematical decision making as students construct the mathematical model that they will use to answer the two questions: How many packages of spaghetti? And at what cost?

## ASSESSING MATHEMATICAL PRACTICE STANDARD 4

In assessing student proficiency with Mathematical Practice Standard 4, modeling with mathematics, there are a number of things to look for:

- Is the student using an appropriate model for the actual situation?
- Is the student sufficiently capable with the mathematics of the model to use it effectively?
- Does the student improve the model if results are examined and found to be somewhat unsatisfactory?
- Does the student recognize when assumptions need to be made?
- Does the student make appropriate assumptions?

## SUMMARY

In order for students to engage in Mathematical Practice Standard 4:

- Teachers must provide contextual problems that can be modeled mathematically, ideally rich problems or at least problems that can be approached in a variety of ways.
- Students must learn to pay attention to and defend the assumptions they make.
- Students must consider their solution in light of the realities of the context and decide if either the solution needs to be tweaked or the assumptions need to be reconsidered.
- Teachers should regularly request justification for a student's choice of model.

• CHAPTER 5 •

# Using Tools Strategically

## *Mathematical Practice Standard 5*

***MP5. Use appropriate tools strategically.***

Mathematically proficient students consider the available tools when solving a mathematical problem. These tools might include pencil and paper, concrete models, a ruler, a protractor, a calculator, a spreadsheet, a computer algebra system, a statistical package, or dynamic geometry software. Proficient students are sufficiently familiar with tools appropriate for their grade or course to make sound decisions about when each of these tools might be helpful, recognizing both the insight to be gained and their limitations. For example, mathematically proficient high school students analyze graphs of functions and solutions generated using a graphing calculator. They detect possible errors by strategically using estimation and other mathematical knowledge. When making mathematical models, they know that technology can enable them to visualize the results of varying assumptions, explore consequences, and compare predictions with data. Mathematically proficient students at various grade levels are able to identify relevant external mathematical resources, such as digital content located on a website, and use them to pose or solve problems. They are able to use technological tools to explore and deepen their understanding of concepts.

**THIS STANDARD** brings to the forefront the issue that the tools we use make a difference in mathematical situations. It is not always about whether a particular tool is a good one in general but whether it is useful in a particular situation.

As teachers, it is important to refrain from telling students what tool to use for a problem in order to give them experience in deciding—in their own opinion—what tool they prefer to employ in a particular situation. There is frequently not a single best tool for any given problem; it depends on student comfort and familiarity with the tool and depends on the student's interpretation of the mathematical situation.

For example, many students would prefer using a virtual manipulative, but others prefer concrete manipulatives. Many students prefer using a number line to do work with integers, but others prefer two-sided counters.

In this chapter I provide examples of problems approached with concrete tools, pictorial tools, and technological tools.

## CONCRETE TOOLS

There are so many valuable concrete tools that it is hard to know where to begin in describing them. Certainly counters, pattern blocks, base ten blocks, and square tiles are included in the list.

Counters are useful, of course, because so much of the concept of number is built on counting.

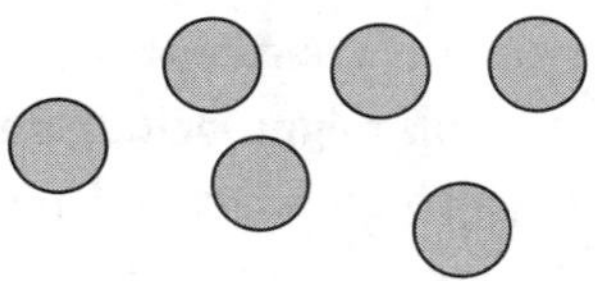

Counters arranged in arrays are particularly useful for bringing out multiplication and division properties.

Pattern blocks are useful in both geometric and numeric situations.

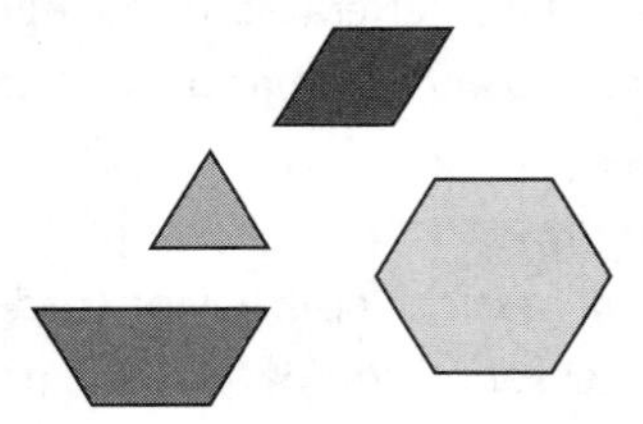

For example, it is possible to use pattern blocks to explore symmetry, tessellations, composing and decomposing shapes, and angle sizes. But it is also possible to use pattern blocks to support proportional reasoning and, in particular, work with fractions.

Base ten blocks are useful for representation of whole numbers and decimals, as well as for support of all of the algorithms, or procedures, for computation with whole numbers and decimals.

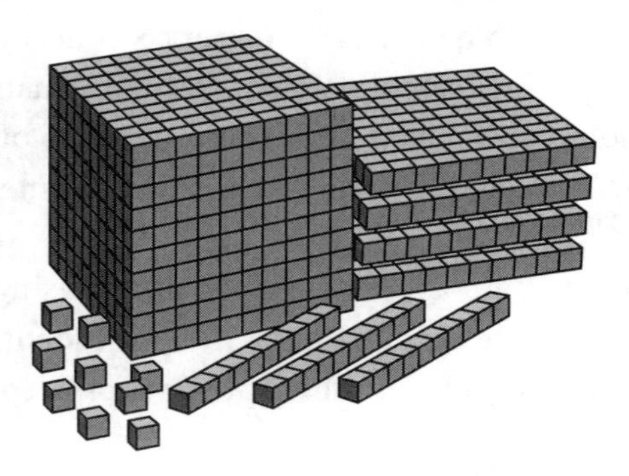

Square tiles are particularly useful. One valuable characteristic of square tiles is that they bridge number and measurement, because students can use them as concrete counters but can also push them together to make continuous shapes such as rectangles.

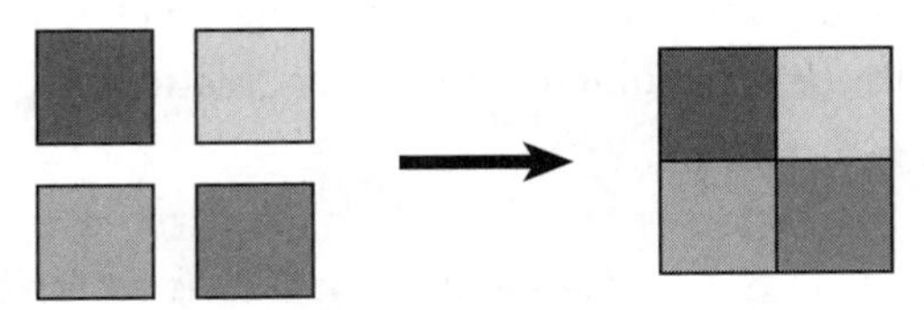

Students are likely to use rulers and protractors on a regular basis in the performance of measurement tasks, but teachers might find it beneficial to have a variety of types of rulers available from which students can choose (e.g., only Imperial measurement, only metric, or both). Teachers might also want to consider making both circular and semicircular protractors available.

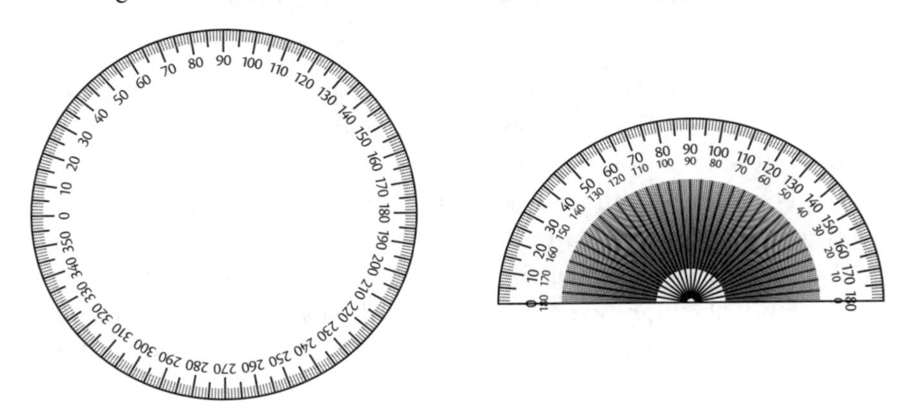

## EXAMPLES OF PROBLEMS INVOLVING CONCRETE TOOLS THAT MIGHT BRING OUT MATHEMATICAL PRACTICE STANDARD 5

### Grades K–2

**Which Tool?**

- You have to add 24 and 8.

? If you didn't already know the answer, what tools would you choose to help you figure it out? Why?

This problem very directly involves students making a choice of tools, whether counters, ten-frames, base ten blocks, or whatever. While many students are likely to choose counters, many others might choose number lines, linking cubes, or ten-frames.

➤ ***Variations.*** A similar approach could be applied to creation of problems involving different computations or (at higher grade levels) the solution of rate, ratio, or percent situations, or problems involving creation of geometric constructions.

**Twice as Many Tens as Ones**

- A number can be broken up into a number of tens and a number of ones.
- There are twice as many tens as ones.

? What could the number be?
What tools could you use to figure out the answer?

Although students might choose to use base ten blocks to represent the mathematics presented in this problem, they might also use place value mats with counters or even links of ten cubes along with single cubes.

For example, students asked to represent 21 might use base ten blocks and choose 2 ten-rods and 1 one-block, or they might use a place value mat and show 2 counters in the tens column and 1 counter in the ones column.

Students might use "standard" numbers like 21, 42, 63, or 84, but there are many other possibilities including 105 (10 tens and 5 ones), 126 (12 tens and 6 ones), 210 (20 tens and 10 ones), and so forth.

**Twenty Blocks**

- You can use any combination of one or more base ten blocks: ones, tens, and hundreds.

? What numbers can you make using EXACTLY 20 blocks?
What do all of your numbers have in common?

Although this problem does not allow for a choice of tools, it does broaden students' understanding of what they can do with base ten blocks beyond very straightforward representations and calculations.

In using their blocks, students discover they can make many numbers, including 20 (20 ones blocks), 29 (1 tens block and 19 ones blocks), 281 (1 hundreds block, 18 tens blocks, and 1 ones block), 200 (20 tens blocks), 668 (6 hundreds blocks, 6 tens blocks, and 8 ones blocks), 776 (7 hundreds blocks, 7 tens blocks, and 6 ones blocks), 983 (9 hundreds blocks, 8 tens blocks, and 3 ones blocks), and 2000 (20 hundreds blocks).

But what do all of the numbers have in common? It turns out that the sum of their digits is either 2, or 11, or 20. Then students can explore why that makes sense.

## Grades 3–5

**Build a Triangle**

Build a triangle with an area half blue.

In this problem, students might initially consider using square tiles but realize that it would be hard to make a "real" triangle, although they might construct something like this (where dark gray blocks represent blue tiles):

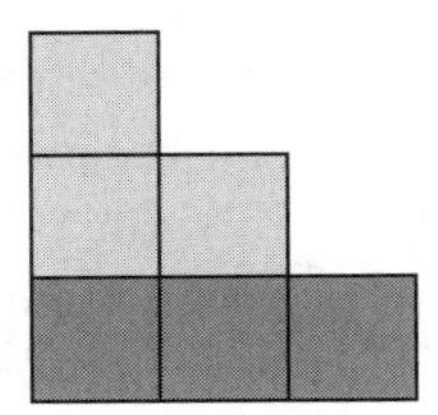

But they might also consider using pattern blocks. Students would have to figure out both how to make triangles with the blocks and how to ensure that half the triangle is blue (i.e., made up of rhombus blocks, which are blue in pattern block sets; shown here in dark gray). The second part of this process (ensuring half is blue) usually demands some use of equivalent fractions. Possible solutions include these:

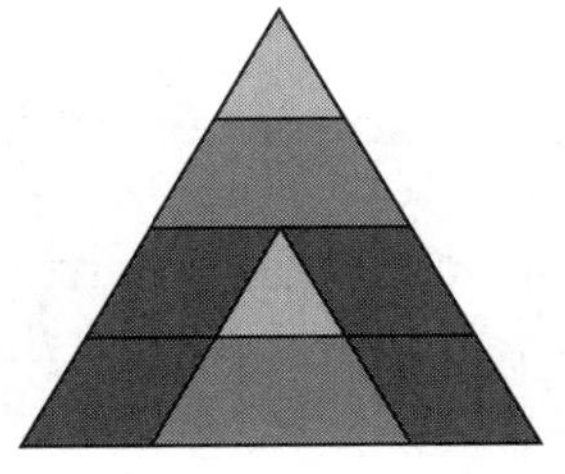

➢ ***Variations.*** Other versions of this problem can be created by changing the fraction one-half to other fractions.

> **Half of a Rectangle**
>
> - You build a rectangle with side lengths that are whole numbers.
> - You cut it in half, based on area.
>
> ? Could the new perimeter be half of the old one?
>
> What fractions of the old perimeter could the new perimeter be?

In this case, students might use rulers or square tiles. They also might use the table feature of a word processing program to create a table of values. If they use tiles, they will discover why the new perimeter is probably greater than one half of the old one.

Look at this example:

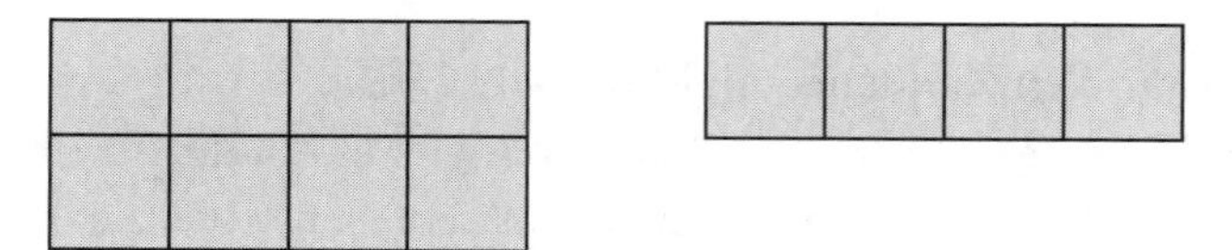

Although the area on the right is half of the area on the left, the perimeter is actually $\frac{10}{12}$, or $\frac{5}{6}$ of the old one. That is because the top and bottom parts of the perimeter are not shrunk at all; only the side pieces are cut in half. Since the side pieces are small compared to the top and bottom, not much perimeter is lost.

The student whose work is shown below needs improvement on Mathematical Practice Standard 6, attending to precision.

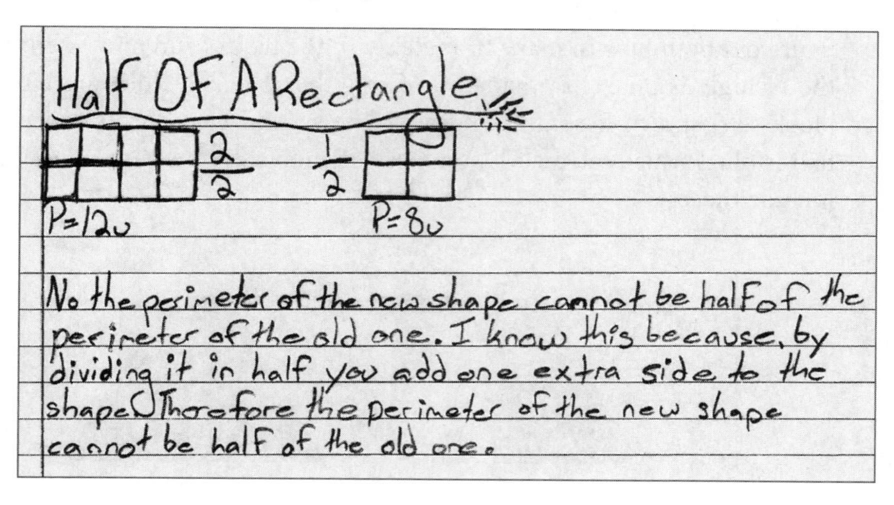

## Grades 6–8

**Mean and Median**

- The mean of a set of five numbers is double the median.

? What could the numbers be?

In working on this problem, students might decide to use linking cubes as a tool. They might, for example, simply choose an even number (since it is a double) as the mean and make 5 identical stacks of cubes showing that number. Because the mean is the result of adding all of the data and dividing by 5, the total of the data set is represented by the total of 5 sets of the mean.

Then the cubes can be redistributed, ensuring that the middle stack is half as tall as the selected mean, and making sure that the first two stacks are no taller than the middle stack. No cubes have been added or removed, so the total number and the mean are unchanged, and the median is half of the mean, as required.

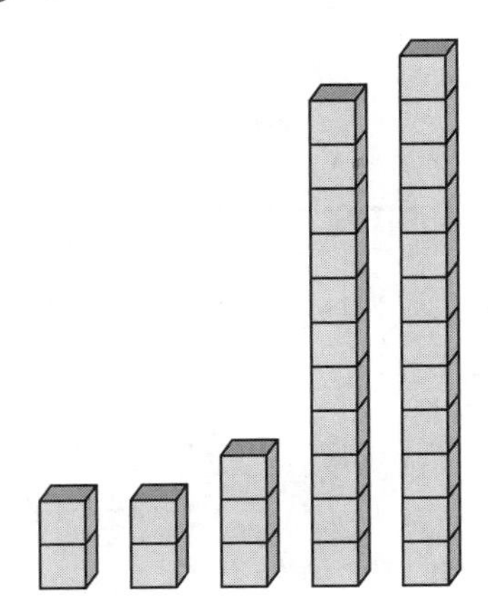

The solution shown above gives a data set of 2, 2, 3, 11, and 12. The median is 3 and the mean is 6. There are, of course, many other solutions.

## PICTORIAL TOOLS

One particularly valuable pictorial tool for students addressing mathematical problems is the number line. In Chapter 4, for example, I presented several situations where the number line was used—even though it was not required—to help solve problems. In one case the problem was for very young children and in another case the problem was for older children. Application of the number line is illustrated in some of the problems below.

## EXAMPLES OF PROBLEMS INVOLVING PICTORIAL TOOLS THAT MIGHT BRING OUT MATHEMATICAL PRACTICE STANDARD 5

### Grades 3–5

**Adding Bunches**

- You add a bunch of 8s to a number and you end up at 156.
- Then you add a bunch of 12s to that same number, and you still end up at 156.

? What was the number, and what did you add each time?

A student might choose to begin at 156 on an open number line and move backward in jumps of 12 and jumps of 8 until the two kinds of jumps arrive at the same place. For example:

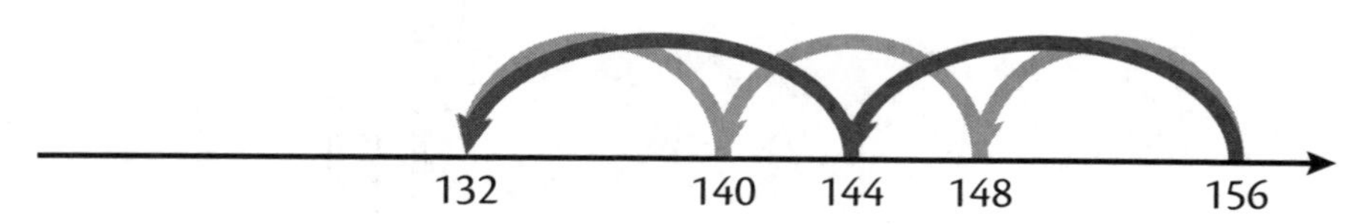


### Grades 6–8

**Twelve Apart**

- You start with the integer *a*.

? What integer *b* could you use so that $a + b$ is 12 LESS than $a - b$? Is there only one possibility?

Using a number line, a student would discover that if $a + b$ is represented as a distance of $b$ beyond $a$ (i.e., to the right of $a$), and $a - b$ as a distance of $b$ to the left of *a,* it becomes obvious that the distance between the difference and the sum is 2*b.*

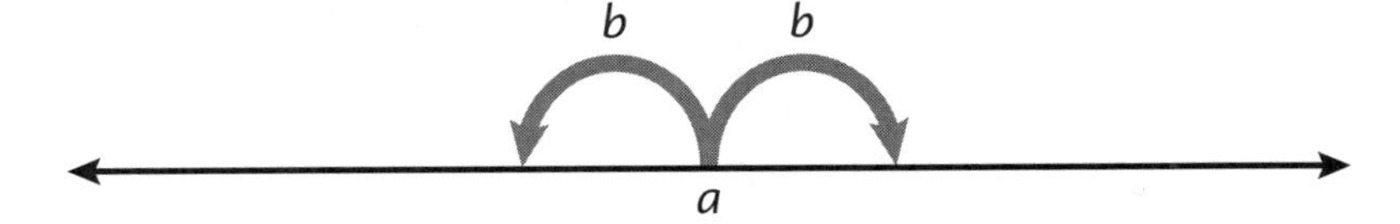


But this assumes that *b* is positive. In fact, if the sum is less than the difference, as required in this problem, then *b* must be negative.

Note how the student whose work is shown below approached the problem:

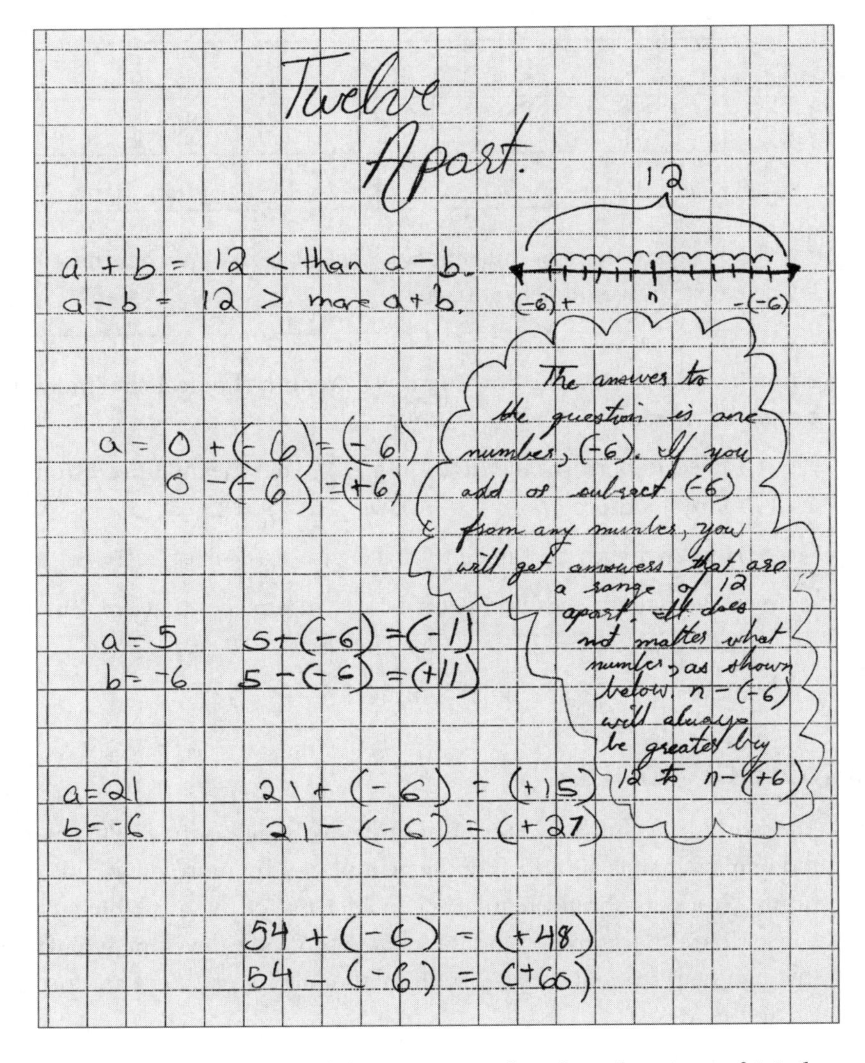


The work shown above is also an example of application of Mathematical Practice Standard 7, recognizing and using structure.

**Fraction Division**

- Notice these fraction rectangles:

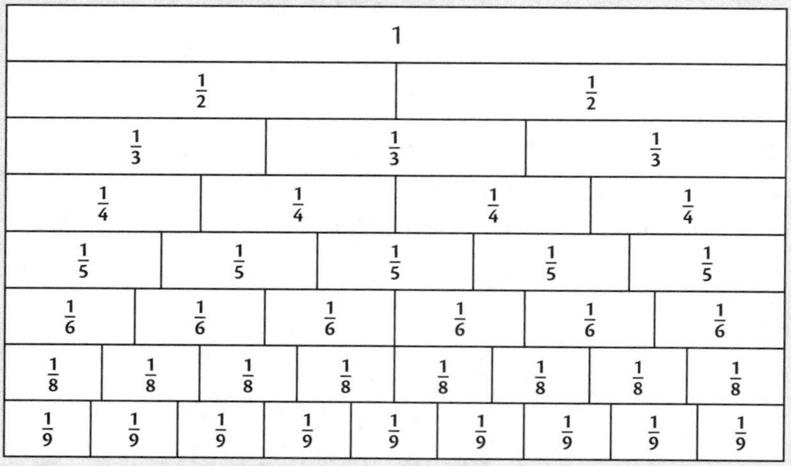

$1$

$\frac{1}{2}$ $\frac{1}{2}$

$\frac{1}{3}$ $\frac{1}{3}$ $\frac{1}{3}$

$\frac{1}{4}$ $\frac{1}{4}$ $\frac{1}{4}$ $\frac{1}{4}$

$\frac{1}{5}$ $\frac{1}{5}$ $\frac{1}{5}$ $\frac{1}{5}$ $\frac{1}{5}$

$\frac{1}{6}$ $\frac{1}{6}$ $\frac{1}{6}$ $\frac{1}{6}$ $\frac{1}{6}$ $\frac{1}{6}$

$\frac{1}{8}$ $\frac{1}{8}$ $\frac{1}{8}$ $\frac{1}{8}$ $\frac{1}{8}$ $\frac{1}{8}$ $\frac{1}{8}$ $\frac{1}{8}$

$\frac{1}{9}$ $\frac{1}{9}$ $\frac{1}{9}$ $\frac{1}{9}$ $\frac{1}{9}$ $\frac{1}{9}$ $\frac{1}{9}$ $\frac{1}{9}$ $\frac{1}{9}$

? Create three or four fraction division questions that these fraction strips would help you solve.

Knowing that $a \div b$ asks how many $b$'s are in $a$ should help students invent questions like these:

- I need to perform the division $\frac{1}{3} \div \frac{1}{6}$. That means I need to figure out how many $\frac{1}{6}$'s there are in $\frac{1}{3}$. *Answer:* $\frac{1}{3} \div \frac{1}{6} = 2$.
- How many $\frac{1}{9}$'s do I need to make $\frac{2}{3}$? *Answer:* $\frac{2}{3} \div \frac{1}{9} = 6$.
- How many pieces do I get if I cut $\frac{3}{4}$ into pieces that are $\frac{3}{8}$ in size? *Answer:* $\frac{3}{4} \div \frac{3}{8} = 2$.
- How many $\frac{5}{8}$'s are in $\frac{3}{4}$? *Answer:* $\frac{3}{4} \div \frac{5}{8} = 1\frac{1}{5}$.

## TECHNOLOGY TOOLS

The world of technology is continuously evolving, so the tools that are named here today may change significantly or be replaced by more powerful tools in the near future. Students should regularly consider use of the available technological tools and prepare themselves to adapt as new tools are developed. Current technology that can help students engage with mathematical practices and processes includes the following:

- Calculators
- Word processing software packages, including drawing functions

- Spreadsheet software packages, especially when exploring repeated reasoning (related to Mathematical Practice Standard 8)
- Graphing software
- Dynamic geometry software, when constructing or measuring shapes
- Virtual manipulatives
- The Internet, as a resource to find information that might be required to solve problems

## EXAMPLES OF PROBLEMS INVOLVING TECHNOLOGY TOOLS THAT MIGHT BRING OUT MATHEMATICAL PRACTICE STANDARD 5

### Grades 3–5

**Putting Together Shapes**

? If you put together a 6-sided shape with a 4-sided shape and neither is inside the other, how many sides might the combined shape have?

Although some students might choose to explore this question with concrete shapes, others would be very happy using virtual manipulatives or dynamic geometry software.

As shown below, virtual manipulatives can be valuable to help students discover that the combined shape could have 8 sides, but it might not:

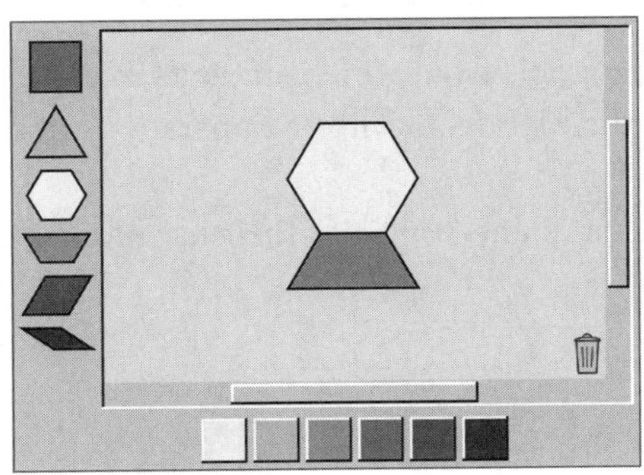
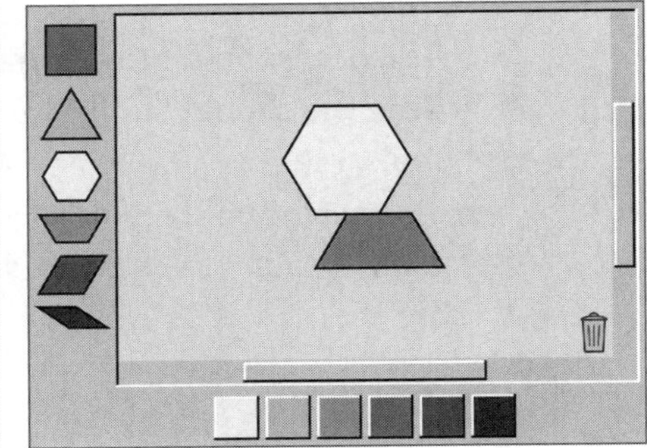

### Grades 6–8

**Population Growth**

? If Canada's growth rate were the same as China's, what would be a good estimate for the Canadian population in 2050?

This problem illustrates a case where the Internet would serve as a valuable resource for gathering critical information as students design an approach for solving a problem. In addition to requiring use of technological tools, this problem also builds on Mathematical Practice Standard 4, modeling with mathematics.

## ASSESSING MATHEMATICAL PRACTICE STANDARD 5

In assessing student proficiency with Mathematical Practice Standard 5, using tools strategically, there are a number of things to look for:

- Is the student considering his or her tool choices?
- Is the student making an appropriate choice of tool?
- Is the student using the tool properly?
- Does the student look up information to help him or her solve mathematical problems?

## SUMMARY

In order for students to engage in Mathematical Practice Standard 5:

- Teachers should provide many tools from which students might choose.
- Teachers should familiarize their students with the operations of these tools.
- Teachers should regularly ask students to justify their use of a particular tool.

• CHAPTER 6 •

# Attending to Precision

## *Mathematical Practice Standard 6*

| ***MP6. Attend to precision.*** |
| --- |
| Mathematically proficient students try to communicate precisely to others. They try to use clear definitions in discussion with others and in their own reasoning. They state the meaning of the symbols they choose, including using the equal sign consistently and appropriately. They are careful about specifying units of measure, and labeling axes to clarify the correspondence with quantities in a problem. They calculate accurately and efficiently, express numerical answers with a degree of precision appropriate for the problem context. In the elementary grades, students give carefully formulated explanations to each other. By the time they reach high school they have learned to examine claims and make explicit use of definitions. |

**THIS PRACTICE** encourages precision in vocabulary, careful use of conventions, and clarity of explanations, as well as appropriate precision in calculations, but a degree of precision that is suited to the age of the student. Generally speaking, precision from students is modeled on precision from teachers. When teachers practice precision, so will students.

### ESTIMATES VS. EXACT ANSWERS

Often, teachers tell students when they want an estimate or when they want an exact answer, but is it important that students regularly make their own decisions about whether an estimate or an exact answer is more appropriate.

Students might be asked, for example, whether they would use an exact answer or an estimate in each of these situations and to explain why they made the choice they did:

- *How long it takes to drive to the mall*
- *How far away from a property line a house is when you are building a fence*
- *The distance from Los Angeles to New York*
- *The area of a living room in a house*
- *The reasonableness of the response of 4127 to the calculation 22 × 58.*

When an exact answer is used, students need to be as accurate as possible in attaining that value.

## CALCULATION EFFICIENCY

Expecting calculation efficiency is appropriate, but only once calculations are understood. It is quite reasonable that younger students would use inefficient strategies that are meaningful to them initially, until more efficient strategies make sense to them.

It is also important to realize that the same procedure might be efficient in one situation but inefficient in another. For example, the standard algorithm for subtraction might be quite efficient for subtracting 312 from 589, but less efficient for subtracting 1 from 300.

## APPROPRIATE MEASUREMENT PRECISION

As students get a bit older, they need to learn about precision in measurement situations. If, for example, they are asked to determine the perimeter of a rectangle where the length was given as 4.1 m and the width as 3.22 m, students need to learn that the perimeter measure should be given to the nearest tenth of a meter, and not the nearest hundredth of a meter, since some of the measurements were given only in tenths of a meter.

## APPROPRIATE USE OF RELATION SIGNS

Students need to learn the meaning and appropriate use of $=$ signs, $>$ signs, and $>$ signs. The $=$ sign is a particular issue.

Students need to learn that the equal sign is used to indicate that the value on the left is equivalent to, or another name for, or has the same value as, the value on the right. So it makes sense to write $4 + 3 = 6 + 1$ since the value of $4 + 3$ is the same as the value of $6 + 1$. It makes sense to write $\frac{2}{4} = \frac{3}{6}$ since $\frac{2}{4}$ is another name for $\frac{3}{6}$. Students need to learn that it is just as correct to write $4 + 3 = 7$ as to write $7 = 4 + 3$.

When trying to indicate that 7 is worth less than 8 or $2x$ is worth more than $x$ (if $x$ is positive), students need to learn that they have a choice of writing either

$$7 < 8 \quad \text{or} \quad 8 > 7$$
$$2x > x \quad \text{or} \quad x < 2x$$

Reversability is important in mathematics.

## APPROPRIATE USE OF TERMINOLOGY

There is value in students using simple and natural language to describe things, but the descriptions, especially as students get older, need to be correct. For example, saying that a triangle is a shape with three points is actually incorrect because it is insufficiently precise. Squares have three points and more. Shapes with sides that are not straight, but that have three vertices, are not triangles. Three points on the same line also do not form a triangle.

There are many strategies for improving student use of vocabulary. One is the regular use of a Frayer model, where students indicate definitions, examples, non-examples, and characteristics of various terms. The Frayer models created by different students or pairs of students can be compared to allow students to decide which definitions seem strongest and why. The example below illustrates a Frayer model for the term "rhombus."

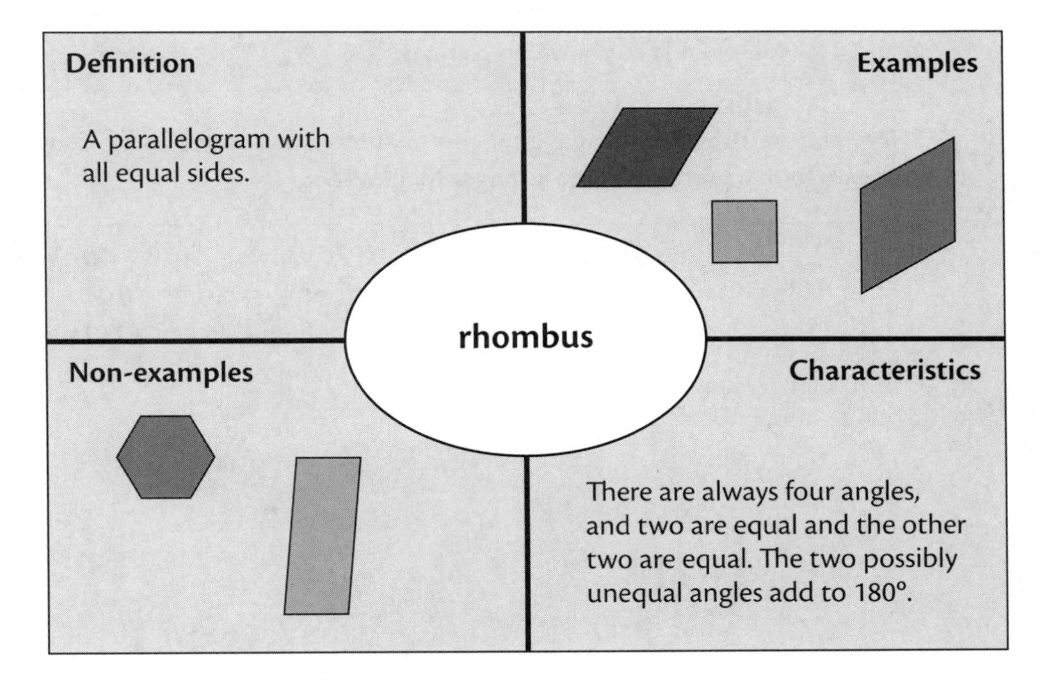


Students might also build other sorts of anchor charts describing what a particular term means.

Another strategy for building vocabulary is to provide a small set of words and ask students to use them in a sentence that makes sense. For example, a sentence that includes the words "rhombus," "square," and "angle" might be: *A square is a special type of rhombus where all four angles are the same size.*

Using hall bulletin boards to hang up definitions that younger and older students use might help students see growth in precision. But, most importantly, students rely on their teachers to model for them the appropriate level of precision required.

Students could be asked to explain terms using both words and pictures. Then students in the class might discuss which explanation they see as more powerful and why. The problem shown below is suitable for students in Grades 3–5. This type of problem can be easily adapted for other grades by substituting appropriate terms to be explained.

**How Would You Explain . . .**

- You want to explain what a prime number is or what a prism is.

? How would you explain it in words?

How would you show it in pictures?

Here and on the next two pages are some interesting examples of students' responses to this problem, some stronger than others:

How would you explain..... Grade 6

words: A prime number is a number that's only factor's are 1 and itself. For example 13, 13 is only a product for 1x13, so it's a prime number, and the factors can't be a decimal.

pictures:

13

1 x 13

The only numbers that multiply to give you 13 is 1 and 13

1✓13÷1=13
2x13÷2=6.5
3x13÷3=4.333
4x13÷4=3.25
5x13÷5=2.6
6x13÷6=2.166
7x13÷7=1857
8x13÷8=1.625
9x13÷9=1.444
10x13÷10=1.3
11x13÷11=1.1818
12x13÷12=1.083

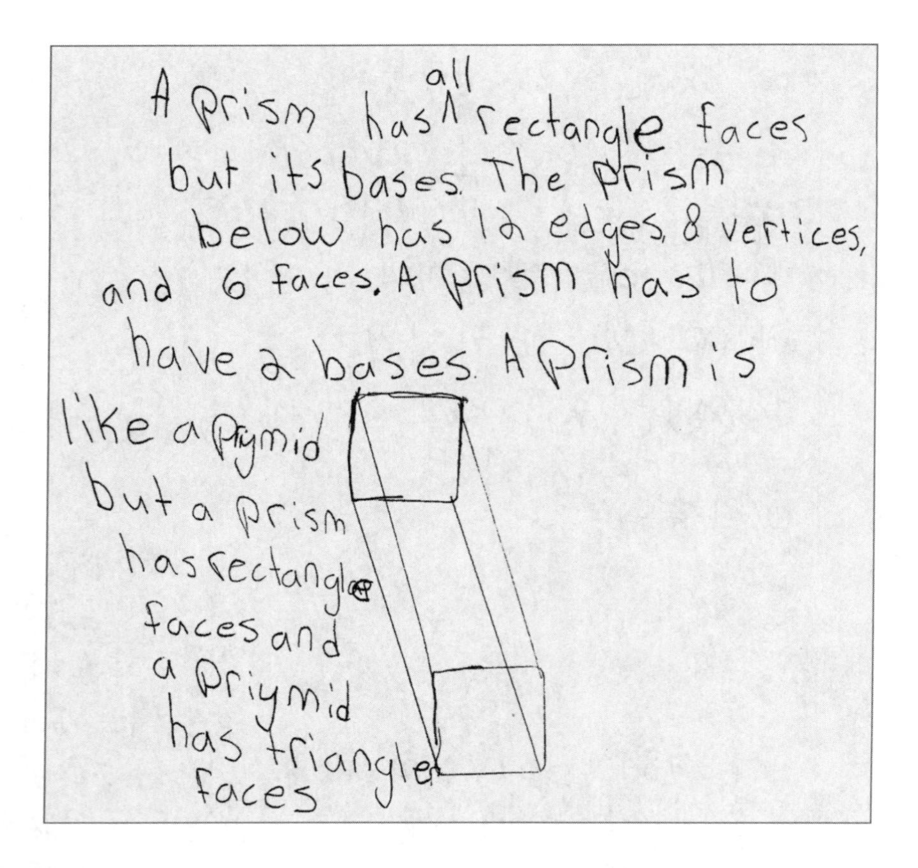


## APPROPRIATE USE OF SYMBOLS FOR VARIABLES

It is important for students to consider carefully whether they are using symbols properly for variables. For example, the equation $x + x = 20$ is very different from the equation $x + y = 20$; in the first case, the two values are identical, and in the second case, they might be identical but might not be. Even if students use symbols rather than letters for unknowns, they should be careful. For example, rather than writing $\square + \square = 20$, they might write $\square + \triangle = 20$ if the values for the two numbers to add to 20 need not be equal.

When students use a variable, it is essential that they indicate what that variable represents. That means that teachers need to regularly ask for this information. For example, if a student solves a problem like *The sum of three consecutive numbers is 51. What were the numbers?* using the equation $x + (x + 1) + (x + 2) = 51$, he or she should be able to respond to each of these questions:

- *Does* x *represent the sum? If not, what does it represent?*
- *What does* x + 2 *represent?*
- *What expression represents the sum of all three numbers?*

## APPROPRIATE USE OF CONVENTIONS IN GRAPHS AND TABLES

Many times data are presented in graphs and tables in mathematical situations. Students need to learn how hard it is to interpret the data unless table rows and columns have proper headings and graphs have proper titles and axis labels.

## APPROPRIATE USE OF UNITS

One of the important student learnings about measurement is that the same measurement can be described in many ways. We can say "1 m" or we can say "100 cm" or we can say "1000 mm." For a variety of reasons, we might choose one rather than another in a particular situation.

This example makes it clear how very important it is to include a unit in a description. If the length of an object were labeled 1000, there would be no sense of its size without the unit accompanying the numerical value.

## ASSESSING MATHEMATICAL PRACTICE STANDARD 6

In assessing student proficiency with Mathematical Practice Standard 6, attending to precision, there are a number of things to look for:

- Is the student correctly defining and interpreting mathematical terms?
- Does the student recognize that the equal sign is used to describe a balance?
- Does the student appropriately label charts and graphs?
- Does the student calculate at an efficiency appropriate for his or her age or grade level?
- Does the student use precision appropriate to the problem or situation and his or her age or grade level?

## SUMMARY

In order for students to engage in Mathematical Practice Standard 6:

- Teachers should regularly include discussions about the level of precision required, or not required, in a particular calculation or measurement situation.
- Teachers should encourage students to use the = sign, the > sign, and the < sign in correct and appropriate ways.
- Teachers should model appropriate levels of precision when using vocabulary, symbols, measurement units, graphs, or tables.

# • CHAPTER 7 •

# Recognizing and Using Structure

## *Mathematical Practice Standard 7*

***MP7. Look for and make use of structure.***

Mathematically proficient students look closely to discern a pattern or structure. Young students, for example, might notice that three and seven more is the same amount as seven and three more, or they may sort a collection of shapes according to how many sides the shapes have. Later, students will see $7 \times 8$ equals the well remembered $7 \times 5 + 7 \times 3$, in preparation for learning about the distributive property. In the expression $x^2 + 9x + 14$, older students can see the 14 as $2 \times 7$ and the 9 as $2 + 7$. They recognize the significance of an existing line in a geometric figure and can use the strategy of drawing an auxiliary line for solving problems. They can also step back for an overview and shift perspective. They can see complicated things, such as some algebraic expressions, as single objects or as being composed of several objects. For example, they can see $5 - 3(x - y)^2$ as 5 minus a positive number times a square and use that to realize that its value cannot be more than 5 for any real numbers $x$ and $y$.

## WHERE DO WE SEE STRUCTURE IN K–8 MATHEMATICS?

Structure comes up in a lot of places in mathematics instruction. In general, structure focuses on generalizations and relationships. For example, there is a structure to what we call fact families, and the same structure applies to any number pair:

***Fact Family***

$2 + 4 = 6$

$4 + 2 = 6$

$6 - 4 = 2$

$6 - 2 = 4$

There is structure associated with number properties, such as the commutative property of addition or multiplication, the associative property of addition or multiplication, or the distributive property of multiplication over addition (or subtraction). These properties lead to important computational strategies such as the following:

***Half and double:*** $a \times b = 2a \times \frac{b}{2}$, so $5 \times 14 = 10 \times 7$

[This is based on the ***associative principle,*** since $a \times 2 \times \frac{b}{2}$ is either $a \times b$ or $(2 \times a) \times \frac{b}{2}$.]

***Shuffle addends:*** $a + b = (a - c) + (b + c)$, so $13 + 28 = 11 + 30$

[This, too, is based on the ***associative principle,*** since $[(a - c) + c] + b$ is either $(a - c) + (b + c)$ or $a + b$.]

***Constant difference:*** $a - b = (a + c) - (b + c)$, so $31 - 18 = 33 - 20$

[This, too, is based on the ***associative principle,*** since $(a + c) + (-c) + (-b)$ is either $(a + c - c) + (-b)$, which is $a - b$, or $(a + c) - (b + c)$, using the ***commutative property*** and the ***zero principle*** as well.]

The ***distributive property*** helps students realize that, for example, $30 + 40 = (3 + 4)$ tens, since $3 \times 10 + 4 \times 10 = 7 \times 10$. Similarly, 3 nickels + 4 nickels = 7 nickels, so $15 + 20 = 35$.

There is structure in various tables with which students work. For example, using a hundred chart, students can see that adding the numbers on the two diagonals of any $2 \times 2$ square on the chart results in the same values.

| 1 | 2 | 3 | 4 | 5 | 6 | 7 | 8 | 9 | 10 |
|---|---|---|---|---|---|---|---|---|---|
| 11 | 12 | 13 | 14 | 15 | 16 | 17 | 18 | 19 | 20 |
| 21 | 22 | 23 | 24 | 25 | 26 | 27 | 28 | 29 | 30 |
| 31 | 32 | 33 | 34 | 35 | 36 | 37 | 38 | 39 | 40 |
| 41 | 42 | 43 | 44 | 45 | 46 | 47 | 48 | 49 | 50 |
| 51 | 52 | 53 | 54 | 55 | 56 | 57 | 58 | 59 | 60 |
| 61 | 62 | 63 | 64 | 65 | 66 | 67 | 68 | 69 | 70 |
| 71 | 72 | 73 | 74 | 75 | 76 | 77 | 78 | 79 | 80 |
| 81 | 82 | 83 | 84 | 85 | 86 | 87 | 88 | 89 | 90 |
| 91 | 92 | 93 | 94 | 95 | 96 | 97 | 98 | 99 | 100 |

This is because the structure of the chart ensures that the four values are

| A | A + 1 |
|---|---|
| A + 10 | A + 11 |

Adding the values on each diagonal results in 2A + 11.

There are similar—and even more—patterns in the addition table or the multiplication table. For example, adding the values on the diagonals of a 2 × 2 square in an addition table means using the terms:

| A + B | A + B + 1 |
|---|---|
| A + 1 + B | A + 1 + B + 1 |

This time, the sum of each of the two diagonals is 2A + 2B + 2.

Multiplying the values on the diagonals of a 2 × 2 square in a multiplication table involves using these terms:

| A × B | A × (B +1) |
|---|---|
| (A + 1) × B | (A + 1) × (B + 1) |

This time, the products of each of the two diagonals is A × B × (A +1) × (B + 1).

Another interesting structural pattern in the multiplication table is that if a term in one row is divided by a term directly below it (or above it) in another row, the value of that quotient remains the same all across the row. For example, each highlighted pair (one number above another) shows a fraction equivalent to $\frac{2}{3}$.

| | 1 | 2 | 3 | 4 | 5 | 6 | 7 | 8 | 9 | 10 |
|---|---|---|---|---|---|---|---|---|---|---|
| 1 | 1 | 2 | 3 | 4 | 5 | 6 | 7 | 8 | 9 | 10 |
| 2 | 2 | 4 | 6 | 8 | 10 | 12 | 14 | 16 | 18 | 20 |
| 3 | 3 | 6 | 9 | 12 | 15 | 18 | 21 | 24 | 27 | 30 |
| 4 | 4 | 8 | 12 | 16 | 20 | 24 | 28 | 32 | 36 | 40 |
| 5 | 5 | 10 | 15 | 20 | 25 | 30 | 35 | 40 | 45 | 50 |
| 6 | 6 | 12 | 18 | 24 | 30 | 36 | 42 | 48 | 54 | 60 |
| 7 | 7 | 14 | 21 | 28 | 35 | 42 | 49 | 56 | 63 | 70 |
| 8 | 8 | 16 | 24 | 32 | 40 | 48 | 56 | 64 | 72 | 80 |
| 9 | 9 | 18 | 27 | 36 | 45 | 54 | 63 | 72 | 81 | 90 |
| 10 | 10 | 20 | 30 | 40 | 50 | 60 | 70 | 80 | 90 | 100 |

There are also structural patterns in the place value system. For example, the reason that the leftmost 5 in 5045 is 1000 times the value of the rightmost 5 is because of the structure of the place value system. The reason that 0.342 is 342 thousandths as well as 34 hundredths + 2 thousandths is based on the structure of the place value system.

There is structure in the way we set up the standard algorithms for addition, subtraction, multiplication, and division. This structure, of course, is rooted in the structure of the place value system.

## HELPING STUDENTS SEE STRUCTURE

One of the ways to help students see structure is to use patterned practice, or strings. For example, asking students to continue this pattern:

$$4 \times 100$$
$$4 \times 10$$
$$4 \times 1$$
$$4 \times 0.1$$

helps explain multiplication of whole numbers by decimals.

Asking students to continue this pattern:

$$7 - 2$$
$$7 - 1$$
$$7 - 0$$
$$7 - (-1)$$

helps explain subtraction of negative integers.

## EXAMPLES OF PROBLEMS THAT MIGHT BRING OUT MATHEMATICAL PRACTICE STANDARD 7

### Grades K–2

**What's Left?**

- You subtract a number from 10.
- What is left is more than what you subtracted.

? What is the most you could have subtracted?

In solving this problem, students might use Mathematical Practice Standard 8 and repeated reasoning to observe that numbers that you could have subtracted are 0, 1, 2, 3, or 4, but no more. But it is structure that would have told the student that the number had to be less than halfway to 10; otherwise, what was left would have been less than what was subtracted.

**No Regrouping**

- Kyla says that you never have to regroup when you subtract.
- She says to just add enough to both numbers so that you don't have to regroup.
- For example, she says that since 51 – 17 has the same result as 54 – 20 (adding 3 to both numbers), she could solve 54 – 20 and not have to regroup.

? Is she right?

Notice that this is also an example of Mathematical Practice Standard 3, constructing and critiquing arguments. Kyla is right because subtraction describes distance between. So the distance between two numbers does not change if you add the same amount to both:

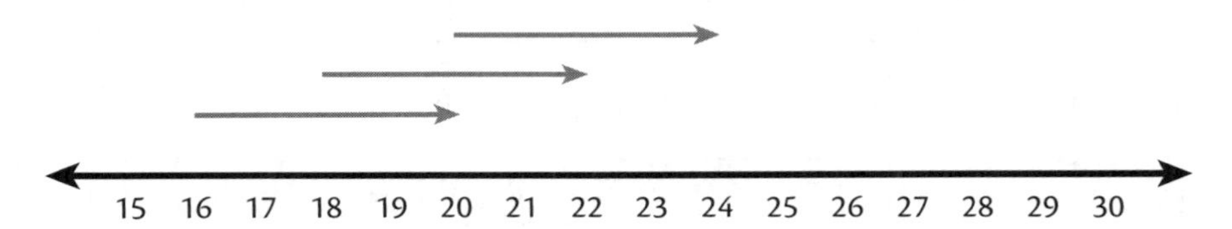


The diagram above shows that $20 - 16 = 22 - 18 = 24 - 20$.

The other piece of structure at play in here is that the digits of numbers form a pattern: 0, 1, 2, . . . , 9, 0, 1, 2 . . . , so regrouping normally is used when a digit of the smaller number is high. By adding, the digits cycle, and digit of the smaller number will eventually become low, so no regrouping will be needed.

## Grades 3–5

**How Do You Know?**

- Alison noticed that when you add a number in the 2 row of the multiplication table to the number in the 5 row below it, you get the number in the 7 row below both of them.

? Why does that work?
Why might that be useful?

Adding rows in a multiplication table, as suggested above, is based on the distributive property of multiplication over addition:

| | 1 | 2 | 3 | 4 | 5 | 6 | 7 | 8 | 9 | 10 |
|---|---|---|---|---|---|---|---|---|---|---|
| 1 | 1 | 2 | 3 | 4 | 5 | 6 | 7 | 8 | 9 | 10 |
| 2 | 2 | 4 | 6 | 8 | 10 | 12 | 14 | 16 | 18 | 20 |
| 3 | 3 | 6 | 9 | 12 | 15 | 18 | 21 | 24 | 27 | 30 |
| 4 | 4 | 8 | 12 | 16 | 20 | 24 | 28 | 32 | 36 | 40 |
| 5 | 5 | 10 | 15 | 20 | 25 | 30 | 35 | 40 | 45 | 50 |
| 6 | 6 | 12 | 18 | 24 | 30 | 36 | 42 | 48 | 54 | 60 |
| 7 | 7 | 14 | 21 | 28 | 35 | 42 | 49 | 56 | 63 | 70 |
| 8 | 8 | 16 | 24 | 32 | 40 | 48 | 56 | 64 | 72 | 80 |
| 9 | 9 | 18 | 27 | 36 | 45 | 54 | 63 | 72 | 81 | 90 |
| 10 | 10 | 20 | 30 | 40 | 50 | 60 | 70 | 80 | 90 | 100 |

One student offered the following explanation:

This would work because when you multiply a numbec to 2, you are adding the number 2 times. When you are multiplying a number to 5, you add the number 5 times. If you multiply a number with 7, you add it 7 times. If you timesed a number 2 times and 5 times, and then added the two numbers, you are really just adding a number 7 times (x7). This would be useful when you don't know how to multiply a high number. You could just break it up and add the numbers together,

(ex: Q= 9 x 5 = ?

Work = 4x5=20 20+25 = 45
5x5=25

**Balance**

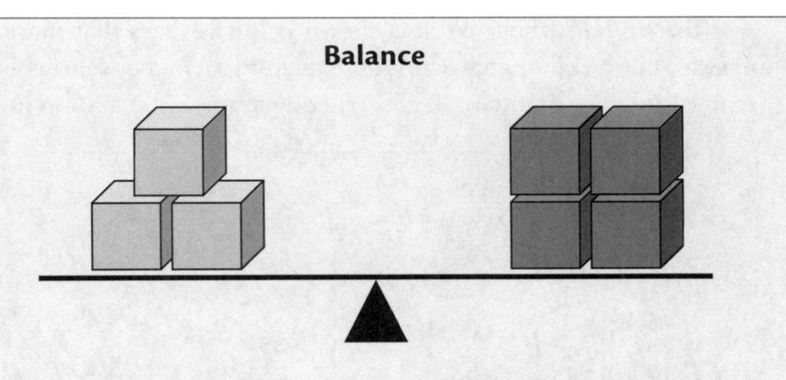

- All the yellow boxes [shown light gray above] hold the same number of metal balls.
- All the red boxes [shown dark gray above] hold the same number of metal balls.
- All of the metal balls are identical.

? How many metal balls might be in each yellow box?

How many might be in each red box?

What is true about ALL possible answers?

In solving this problem, students definitely use the structure of the number system. If a high school student were to represent this problem, he or she might write the equation $3x = 4y$ (where $x$ represents the number of metal balls in a yellow box and $y$ represents the number of metal balls in a red box). Working with that equation, it would become clear that $y = \frac{3x}{4}$, so that the number of balls in a red box must be $\frac{3}{4}$ of the number in a yellow box.

But a much younger student can also solve this problem by reasoning. He or she might realize that the number in a red box must be less than the number in a yellow box, or there would be more weight on the right. He or she might realize that the number in a red box could not be half the number in a yellow box or the total would be the amount in 2 yellow boxes, not 3. He or she might then try values and notice that $3 \times 4 = 4 \times 3$, so 4 in a yellow box and 3 in a red box work.

But since 4 balls in a yellow box and 3 balls in a red box work, so would double those amounts (3 yellow boxes of 8 balls each and 4 red boxes of 6 balls each, for $3 \times 8 = 4 \times 6$) or triple those amounts ($3 \times 12 = 4 \times 9$). Soon, it becomes clear that whatever multiple of 4 is used for a yellow box, that same multiple of 3 must be used for a red box.

The student whose work is shown below realizes that there is more than one answer, but he still needs work on explaining why he went where he did (Mathematical Practice Standard 3, constructing arguments) and on how to generalize:

SO the red dox has more because thar is less red boxes. So thar 4 in each thar is 3 in each box So the red boxes has 12 and the yellow boxes 3 in each that = 12. and the yellow Boxes can Be 6 and the red Boxes can Be 8 Because they Both land ON 24.

Another student comes to a generalization, but needs work on precision, Mathematical Practice Standard 6, since the explanation does not indicate that the same multiple must be used both times:

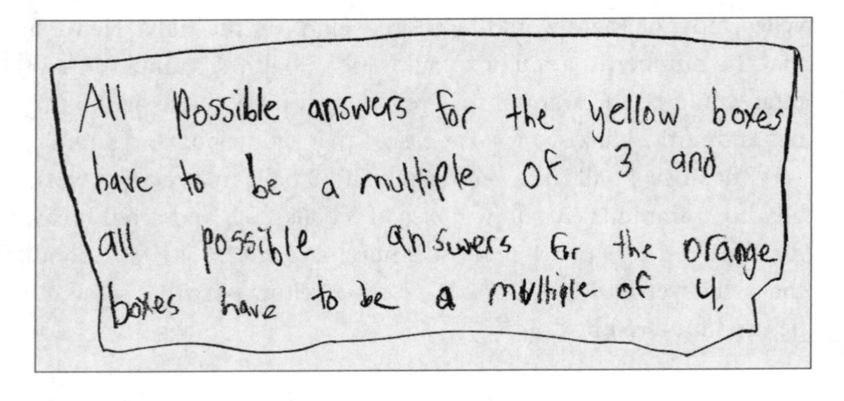
All possible answers for the yellow boxes have to be a multiple of 3 and all possible answers for the orange boxes have to be a multiple of 4.

Finally, the student whose work is shown below indicates the required multiplicative relationship using ratio notation. She indicates that the ratio of the number of balls in a yellow box to the number in a red box is 4:3, even though she does not list any actual values:

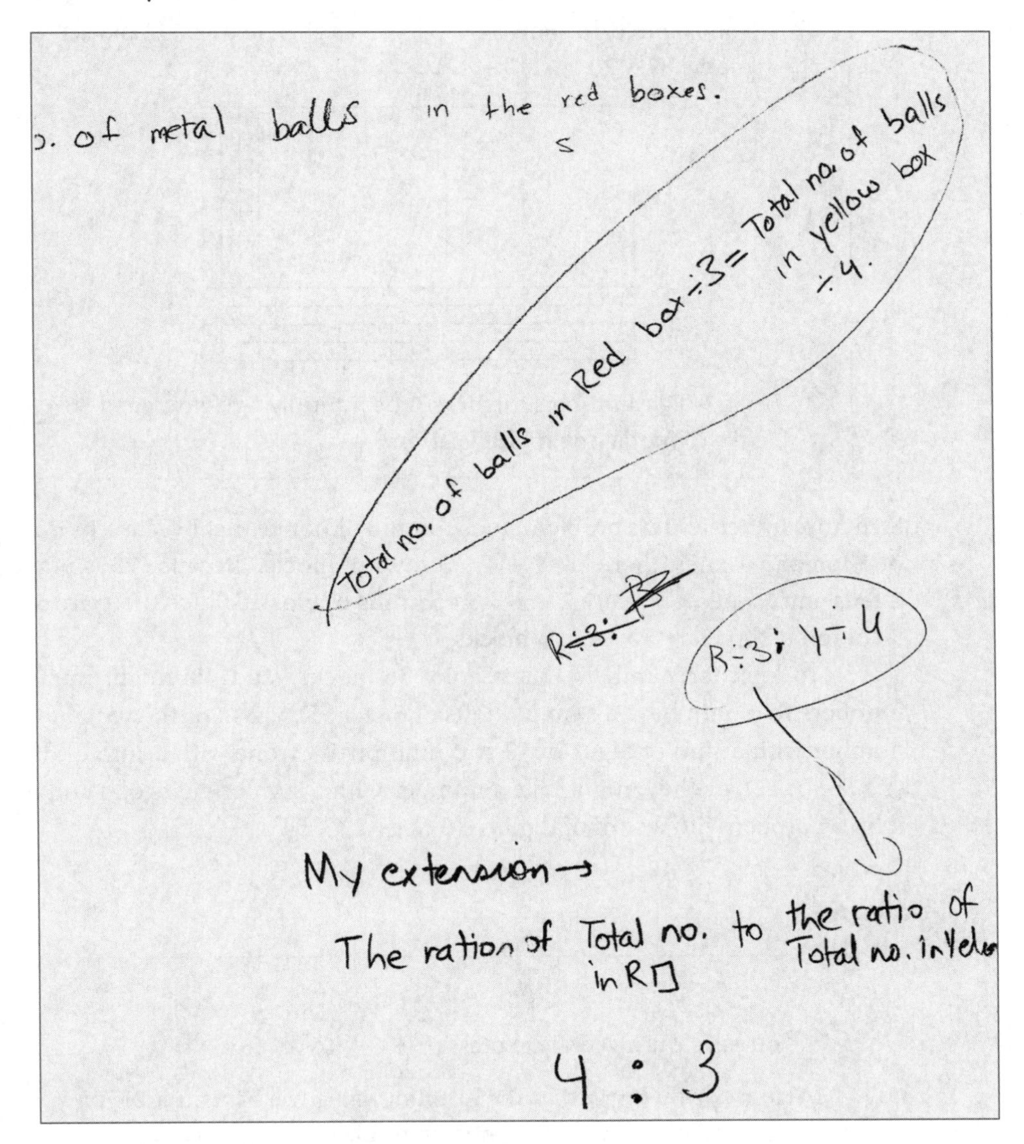


A teacher might prompt the third student for the missing values by asking the following:

*I understand the ratio idea. Can you tell me how that would give actual values?*

**36 Blocks**

- You multiply two 2-digit numbers by using base ten blocks arranged in an array.

  For example, to show 14 × 23, you might use these 25 blocks to model 322 (200 + 110 + 12).

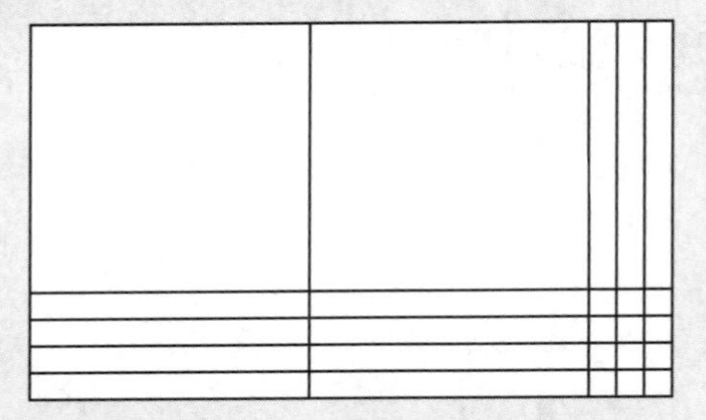

? What 2-digit numbers might you be multiplying if you need 36 blocks to model the multiplication?

Structure underlies this problem, too. You might notice that because 14 is made up of 1 ten and 4 ones, there are 1 + 4 = 5 rows of blocks. Because 23 is made up of 2 tens and 3 ones, there are 2 + 3 = 5 columns of blocks. Since there are 5 rows of 5 columns, there are 5 × 5 = 25 blocks.

So to look for numbers that require 36 blocks, students might multiply two numbers that both have a sum of digits of 6 (e.g., 51 × 24), or they might choose a number with a sum of digits of 12 and multiply it by one with a sum of digits of 3 (e.g., 66 × 12), or they might take a number with a sum of digits of 4 and multiply it by a number with a sum of digits of 9 (e.g., 22 × 54).

## Grades 6–8

**All Fives**

- You have the algebraic expression $5x - 10x^2 + 15x$.

? How do you know that the resulting value will be a multiple of 5, no matter what whole number you substitute for $x$ into the expression?

In responding to this question, students must realize that multiplying a whole number by 5, 10, or 15 results in a multiple of 5; this means it can be represented as groups of 5. Adding or subtracting groups of 5 also results in groups of 5.

➤ ***Variations.*** Similar problems could be created using other algebraic expressions. For example, when substituting a whole number into $4x - 8x^2 + 12$, the result must be a multiple of 4.

**The Tip**

- You often add 15% to the price of a meal in a restaurant for a tip.

? What percent of what you pay (including the tip) is the tip?

Many students will assume that the answer is obvious, that the tip is 15% of the full price, but that is not the case. Suppose B represents the cost of the meal without the tip and A represents the cost of the meal with the tip. Then A = 115% of B.

If the number A is 115% of the number B, then $\frac{A}{B} = \frac{115}{100}$. That means that $\frac{B}{A}$, or the meal's price (without tip) in relation to the total amount paid (tip included) is $\frac{100}{115}$. That turns out to be 0.86956 . . . , or 87%. So 87% of what you paid is the cost of the meal without the tip. That leaves only 13% for the tip.

The interesting question is why a 15% tip does not turn out to be 15% of the final cost. This is an important discussion. It is because the tip amount is 15% of a smaller amount, so the actual final percent for the tip is less than 15% of a larger amount.

**Pattern Rules**

? How do you know that the pattern rule for each of these patterns will involve the expression $3n$?

4, 7, 10, 13, 16, . . .

12, 15, 18, 21, . . .

2, 5, 8, 11, . . .

Students might realize that every pattern that increases in a constant way by the value of 3 has the same structure as 3, 6, 9, 12, . . . , but might be shifted up or down by a given amount. For example, the last pattern is the 3, 6, 9, 12, . . . pattern shifted down by 1, so the rule must be $3n - 1$; the second pattern is the 3, 6, 9, 12, . . . pattern shifted up by 9, so its rule must be $3n + 9$.

**The Handshake Problem**

- In a group of 100 people, each pair will shake hands once.

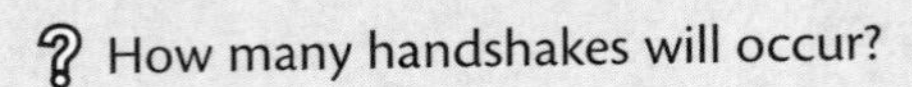

? How many handshakes will occur?

This problem, too, builds on structure. A student might use Mathematical Practice Standard 4 and appeal to a mathematical model.

He or she might think of a circle with 100 dots, each dot representing one person. Each person on that circle steps forward and shakes 99 hands. So since this happens 100 times, there are 9900 handshakes. BUT if that happened, each pair of people shook hands twice, once when the first person stepped forward and once when the second person stepped forward. So there were really only half as many handshakes as 9900, that is, $9900 \div 2 = 4950$ handshakes.

This solution is based on structure, since the model pertains no matter how many people might be involved in the problem.

## ASSESSING MATHEMATICAL PRACTICE STANDARD 7

In assessing student proficiency with Mathematical Practice Standard 7, recognizing and using structure, there are a number of things to look for:

- Does the student observe generalizations that are appropriate for his or her grade level (e.g., that evens + evens = evens, or that adding the numerators and denominators of equivalent fractions results in an equivalent fraction, or that if you save $x$% on a sale, you pay $(100 - x)$%)?
- Does the student not only look for patterns but also look at why the patterns make sense?

## SUMMARY

In order for students to engage in Mathematical Practice Standard 7:

- Teachers should provide problems that lead to generalizations.
- Teachers should often be asking, *Does that happen all the time or just with these numbers?*

• CHAPTER 8 •

# Recognizing and Using Regularity in Repeated Reasoning

## *Mathematical Practice Standard 8*

***MP8. Look for and express regularity in repeated reasoning.***

Mathematically proficient students notice if calculations are repeated, and look both for general methods and for shortcuts. Upper elementary students might notice when dividing 25 by 11 that they are repeating the same calculations over and over again, and conclude they have a repeating decimal. By paying attention to the calculation of slope as they repeatedly check whether points are on the line through (1,2) with slope 3, middle school students might abstract the equation $\frac{y-2}{x-1} = 3$. Noticing the regularity in the way terms cancel when expanding $(x - 1)(x + 1)$, $(x - 1)(x^2 + x + 1)$, and $(x - 1)(x^3 + x^2 + x + 1)$ might lead them to the general formula for the sum of a geometric series. As they work to solve a problem, mathematically proficient students maintain oversight of the process, while attending to the details. They continually evaluate the reasonableness of their intermediate results.

## WHERE MIGHT REPEATED REASONING OCCUR?

Repeated reasoning occurs in mathematical situations for younger students as well as in situations more appropriate for older students.

For example, very young students recognize that every time you add 1, you say the next number. Slightly older students observe that when you add a 2-digit number to its "reverse" number, for example, 32 to 23 or 92 to 29, you usually get a palindrome (a number that reads the same forward as backward), but not always, since 91 + 19 = 110.

Young students notice the patterns in how we say numbers; this helps them deal with numbers they have never met before. For example, they know that the number after 429 must be 430 since that is the pattern they learned.

Older students notice that on a fraction tower, all of the equivalents to $\frac{1}{2}$ have a numerator half the size of the denominator:

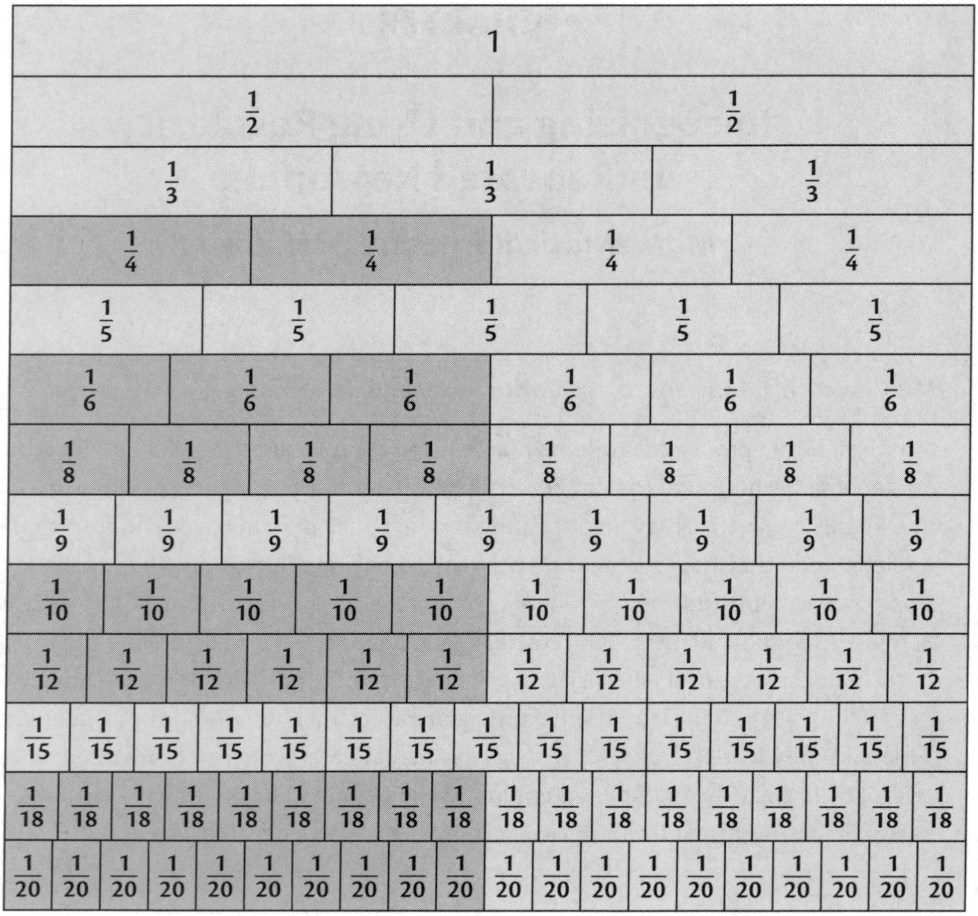

1

$\frac{1}{2}$ $\frac{1}{2}$

$\frac{1}{3}$ $\frac{1}{3}$ $\frac{1}{3}$

$\frac{1}{4}$ $\frac{1}{4}$ $\frac{1}{4}$ $\frac{1}{4}$

$\frac{1}{5}$ $\frac{1}{5}$ $\frac{1}{5}$ $\frac{1}{5}$ $\frac{1}{5}$

$\frac{1}{6}$ $\frac{1}{6}$ $\frac{1}{6}$ $\frac{1}{6}$ $\frac{1}{6}$ $\frac{1}{6}$

$\frac{1}{8}$ $\frac{1}{8}$ $\frac{1}{8}$ $\frac{1}{8}$ $\frac{1}{8}$ $\frac{1}{8}$ $\frac{1}{8}$ $\frac{1}{8}$

$\frac{1}{9}$ $\frac{1}{9}$ $\frac{1}{9}$ $\frac{1}{9}$ $\frac{1}{9}$ $\frac{1}{9}$ $\frac{1}{9}$ $\frac{1}{9}$ $\frac{1}{9}$

$\frac{1}{10}$ $\frac{1}{10}$ $\frac{1}{10}$ $\frac{1}{10}$ $\frac{1}{10}$ $\frac{1}{10}$ $\frac{1}{10}$ $\frac{1}{10}$ $\frac{1}{10}$ $\frac{1}{10}$

$\frac{1}{12}$ $\frac{1}{12}$ $\frac{1}{12}$ $\frac{1}{12}$ $\frac{1}{12}$ $\frac{1}{12}$ $\frac{1}{12}$ $\frac{1}{12}$ $\frac{1}{12}$ $\frac{1}{12}$ $\frac{1}{12}$ $\frac{1}{12}$

$\frac{1}{15}$ $\frac{1}{15}$ $\frac{1}{15}$ $\frac{1}{15}$ $\frac{1}{15}$ $\frac{1}{15}$ $\frac{1}{15}$ $\frac{1}{15}$ $\frac{1}{15}$ $\frac{1}{15}$ $\frac{1}{15}$ $\frac{1}{15}$ $\frac{1}{15}$ $\frac{1}{15}$ $\frac{1}{15}$

$\frac{1}{18}$ $\frac{1}{18}$ $\frac{1}{18}$ $\frac{1}{18}$ $\frac{1}{18}$ $\frac{1}{18}$ $\frac{1}{18}$ $\frac{1}{18}$ $\frac{1}{18}$ $\frac{1}{18}$ $\frac{1}{18}$ $\frac{1}{18}$ $\frac{1}{18}$ $\frac{1}{18}$ $\frac{1}{18}$ $\frac{1}{18}$ $\frac{1}{18}$ $\frac{1}{18}$

$\frac{1}{20}$ $\frac{1}{20}$ $\frac{1}{20}$ $\frac{1}{20}$ $\frac{1}{20}$ $\frac{1}{20}$ $\frac{1}{20}$ $\frac{1}{20}$ $\frac{1}{20}$ $\frac{1}{20}$ $\frac{1}{20}$ $\frac{1}{20}$ $\frac{1}{20}$ $\frac{1}{20}$ $\frac{1}{20}$ $\frac{1}{20}$ $\frac{1}{20}$ $\frac{1}{20}$ $\frac{1}{20}$ $\frac{1}{20}$

Some 3rd-grade students noticed the following when they did some subtractions:

$$\begin{array}{r} 41 \\ -\ 28 \\ \hline 20 - 7 \end{array} \qquad \begin{array}{r} 52 \\ -\ 17 \\ \hline 40 - 5 \end{array} \qquad \begin{array}{r} 85 \\ -\ 28 \\ \hline 60 - 3 \end{array}$$

They observed:

> *You can subtract the tens the normal way. You can subtract the ones "upside down" and subtract that value from the tens. You get the right answer that way. They then wondered if this always works.*

This is certainly a case of students using repeated reasoning to make a conjecture. It turns out that they were correct.

Older students might notice that every 3rd multiple of 4 is also a multiple of 3 (4, 8, **12**, 16, 20, **24**, 28, 32, **36**, . . .) or that every term in the pattern 8, 11, 14, 17,

20, . . . is exactly 2 more than a multiple of 3. Or they might observe that whenever they substitute whole number values for $x$ in the expression $3x - 9x^2$, they get multiples of 3. They can explore the whys of the underlying mathematics by using Mathematical Practice Standard 7, recognizing and using structure.

## EXAMPLES OF PROBLEMS THAT MIGHT BRING OUT MATHEMATICAL PRACTICE STANDARD 8

### Grades K–2

**Forward and Back**

- You are on the number 5 on this path.

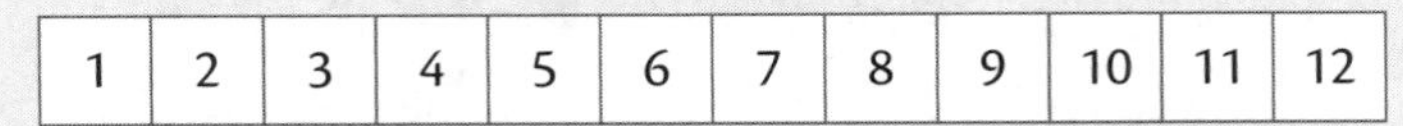


- You move SOME steps forward.
- Then you move SOME steps back.
- You repeat both moves, with exactly the same numbers of steps as the first time.
- You land on 9.

? How many steps forward might you have gone and how many steps back?

Although older students might analyze the problem to figure out what could happen, younger students are likely to guess and test. Initially, most students will simply move 4 forward and may need reminding that they have to go back. At some point, students might then move 5 forward and go 1 back; this would land them at 9, but they have forgotten the need for repeating the moves.

The student whose work is shown at the top of the next page has managed to get to 9 but has misapplied the rules in yet a different way:

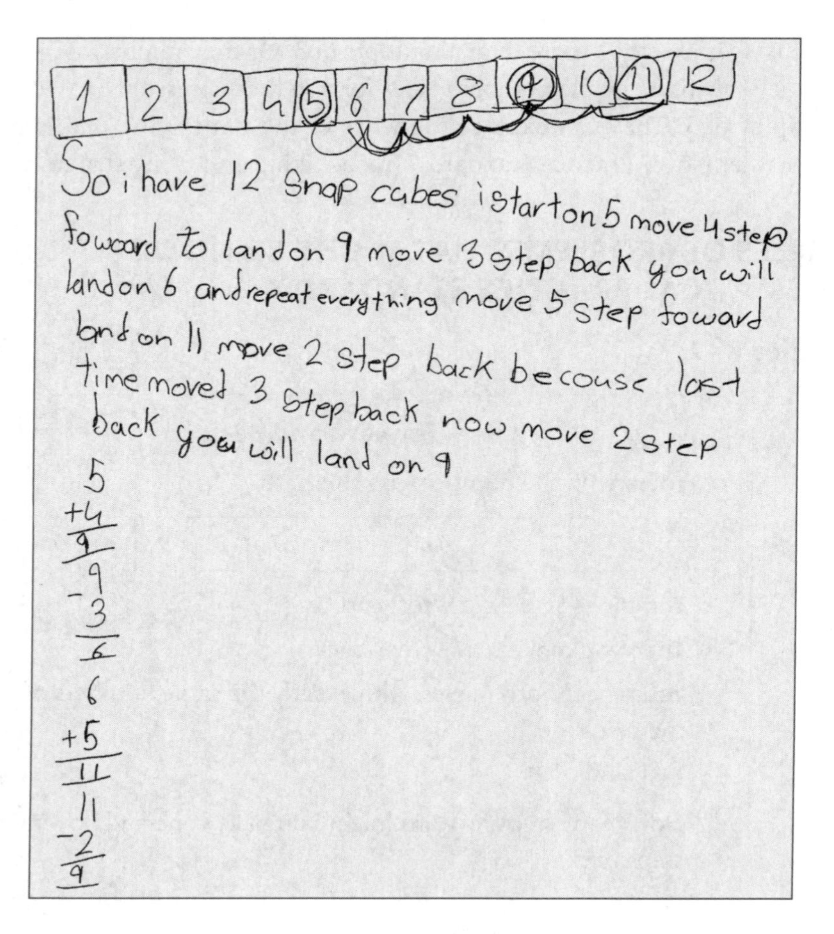


This individual has taken the same total number of steps in each set of forward and backward moves, but the two forward moves differ from each other and the two backward moves differ from each other.

After more trials, students might realize that you could go forward 3 and back 1 each time (landing on 8, then 7, then 10, then 9) or forward 4 and back 2 each time (landing on 9, 7, 11, 9), and so forth.

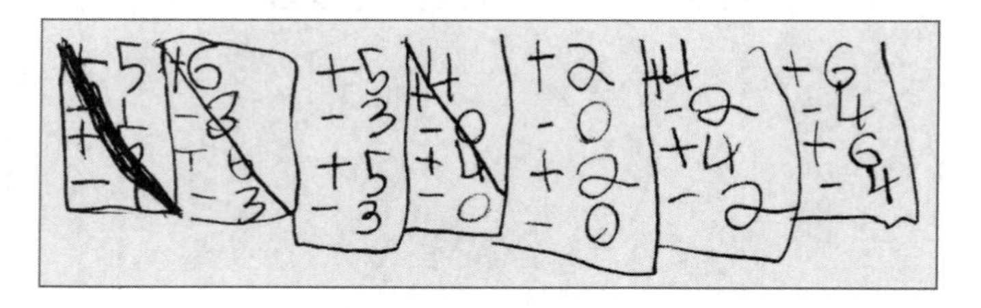


Older students might utilize Mathematical Practice Standard 7, recognizing and using structure, to see that if it takes 2 identical moves to move 4, each move must have the effect of moving forward 2, and that if a forward minus backward move is 2, all that matters is that the forward move is 2 more than the backward move.

**Reversing Digits**

- You choose a two-digit number.
- You reverse the order of the digits.
- You subtract the smaller number from the greater one.
- The answer is 36.

? What could the pair of numbers be?
What is true about all of the possible results?

As students explore what happens when you subtract the reverse of a two-digit number from the original number, they might observe that the result is always a multiple of 9. For example:

$$\begin{array}{r}83\\-38\\\hline45\end{array}\qquad\begin{array}{r}54\\-45\\\hline9\end{array}\qquad\begin{array}{r}72\\-27\\\hline45\end{array}\qquad\begin{array}{r}61\\-16\\\hline45\end{array}$$

And the multiple of 9 that it is happens to be the difference between the two digits. So if the difference is 36, which is $4 \times 9$, the difference between the digits is 4. In this case, possible numbers are 51 – 15, 62 – 26, 73 – 37, 84 – 48, or 95 – 59.

## Grades 3–5

**Square Numbers**

- Josh decided to add consecutive sequences of odd numbers. He noticed that

$$1 + 3 = 4$$
$$1 + 3 + 5 = 9$$
$$1 + 3 + 5 + 7 = 16$$
$$1 + 3 + 5 + 7 + 9 = 25$$

- He decided that the answer is always a square number.
- He predicts that this will always be true.

? Do you agree or disagree? How can you be sure?

First of all, students need to notice that the results of the additions are all square numbers; many students don't look for generalizations, but Josh, above, did. Most students will try one or maybe two more examples and might be convinced.

But the interesting question is, *How can you be sure?* At this point, students are likely to bring in other Mathematical Practice Standards (e.g., Standard 1, making sense of problems and persevering in solving them). They might use visual tools to help see why Josh's prediction could be true. The diagram below should help make the reason clear:

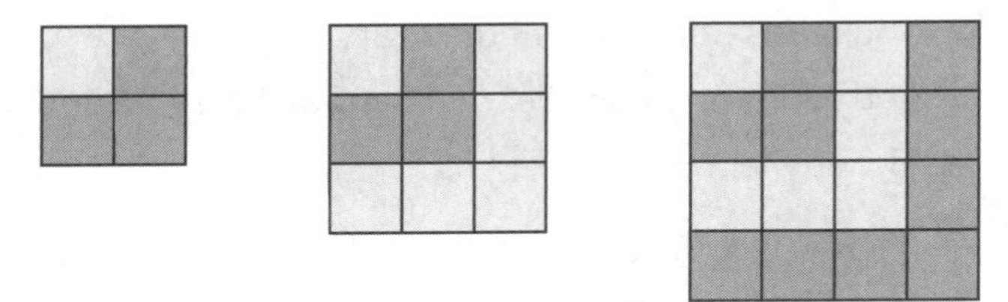

**Left Over**

- You have some counters.
- When you create groups of 3, there is 1 counter left over.
- When you create groups of 4, there are 3 counters left over.

? How many counters might you have?
What do you notice about the relationship between the possible answers?

Students exploring this problem might notice that the number of counters could be 7, 19, 31, 43, 55, 67, . . . . They might then realize that perhaps if you keep adding 12, you get other answers.

It turns out that they are correct. Again, the interesting question becomes, *Why?* This brings up Mathematical Practice Standard 7, recognizing and using structure, because it is structure that tells us why adding 12 works. Notice that 12 is a common multiple of 3 and 4, and therefore adding a group of 12 has no impact on the remainders when creating groups of 3 and 4. Since the number 7 works, so will 19 (there are more groups of 3 and 4 but no change in the remainders).

**Perimeter vs. Area**

- Compare the values (in inches and square inches) of the perimeter and area of each of these rectangles:

  $4 \times 2$ $8 \times 2$ $11 \times 2$ $17 \times 2$ $23 \times 2$

? How do the perimeter and area values compare?
Why does that make sense?

Trying the suggested examples, students should see that whenever the dimensions of a rectangle are $a \times 2$, the perimeter value, in inches, is always 4 more than the area value in square inches.

Students might use Mathematical Practice Standard 2, reasoning abstractly and quantitatively, to represent this situation algebraically and see that the perimeter is $2a + 4$ and the area is $2a$.

**42 Apart**

- The numerator and denominator of a fraction equivalent to $\frac{2}{5}$ are 42 apart.

? What is that equivalent fraction?

As students create equivalent fractions for $\frac{2}{5}$, specifically $\frac{4}{10}, \frac{6}{15}, \frac{8}{20}$, and so forth, they should observe that the differences between numerator and denominator are 3 (for $\frac{2}{5}$), then 6, then 9, then 12, . . . . Students should notice that these numbers increase by 3, but that the differences are also all multiples of 3. Students could conclude that since 42 is the 14th multiple of 3 (beginning at $1 \times 3$), then the fraction is

$$\frac{14 \times 2}{14 \times 5}, \text{ which is } \frac{28}{70}.$$

This observation could turn into an example of the use of Mathematical Practice Standard 7, recognizing and using structure, since this is, in fact, an example of the use of the distributive principle of multiplication over addition. This is because in the fraction

$$\frac{n \times a}{n \times b}$$

the difference between numerator and denominator is $n \times b - n \times a = n(b - a)$.

## Grades 6–8

**$y = mx + m$**

- You graph lines of the form $y = mx + m$ for many different values of m.

? What do you notice about the lines?

Why do you think that happened?

As students do a few graphs—for example, $y = x + 1$, $y = 2x + 2$, and $y = 3x + 3$—they might notice that all three lines go through (−1,0).

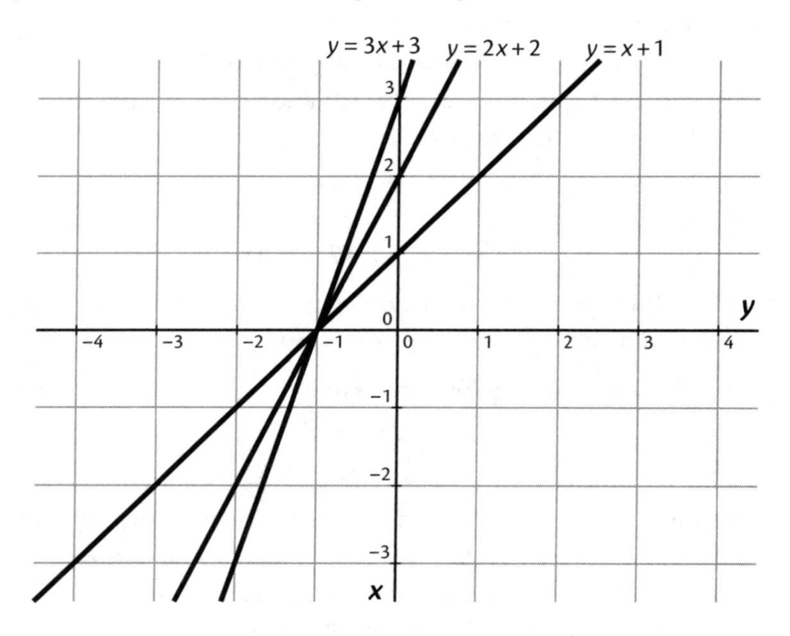


Students might conjecture, using Mathematical Practice Standard 1, making sense of problems and persevering in solving them, that all such lines go through (−1,0). Students might realize that, indeed, if $x = -1$, then $y = -m + m = 0$, so the conjecture is true.

**Greatest Common Factor**

- You graph $y$ = GCF(3,$x$) for different whole number values of $x$.
- You do the same for these equations:

  $y$ = GCF(2,$x$)  $y$ = GCF(5,$x$)  $y$ = GCF(7,$x$)

- Next you graph $y$ = GCF(4,$x$) for different whole number values of $x$.

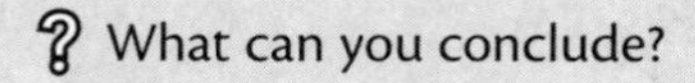
What can you conclude?

The five graphs requested are shown on the next two pages:

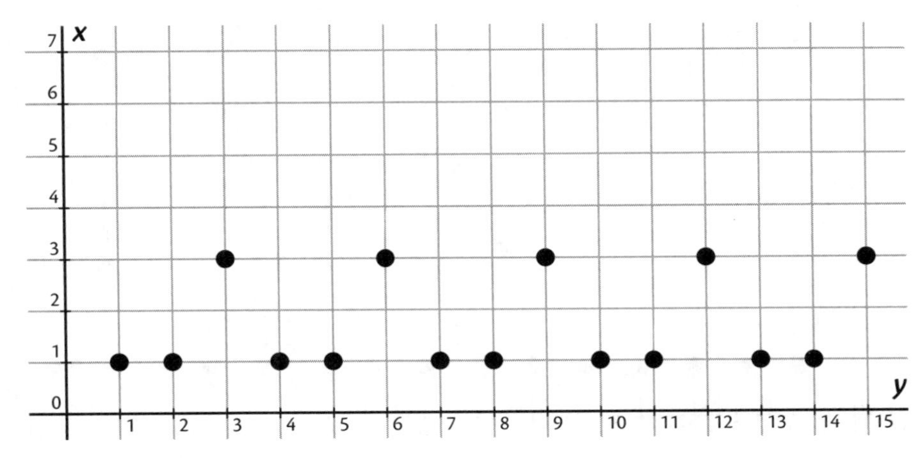


$y = \text{GCF}(2,x)$

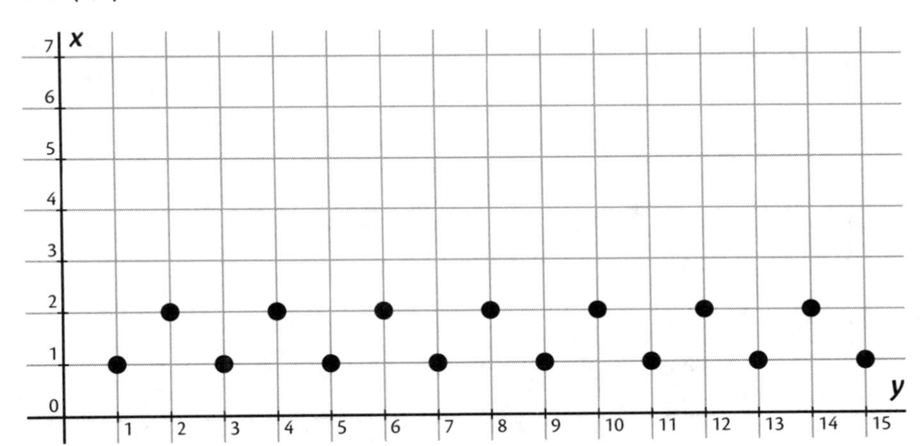


$y = \text{GCF}(5,x)$

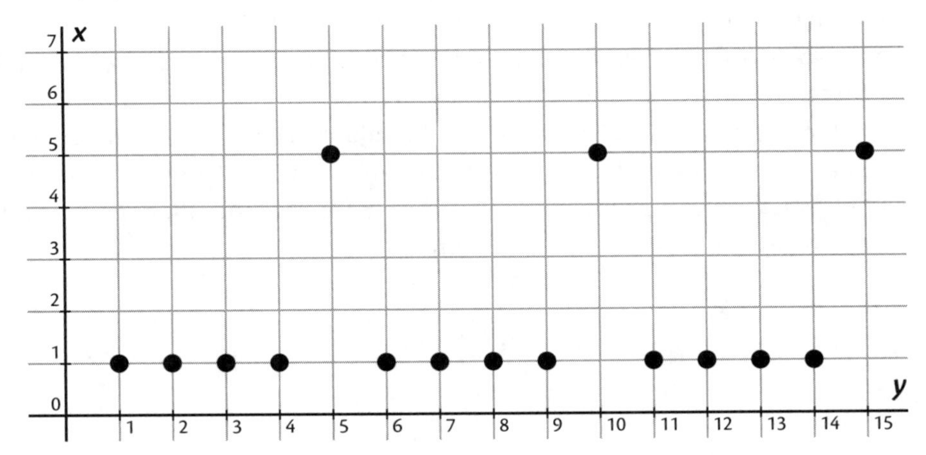

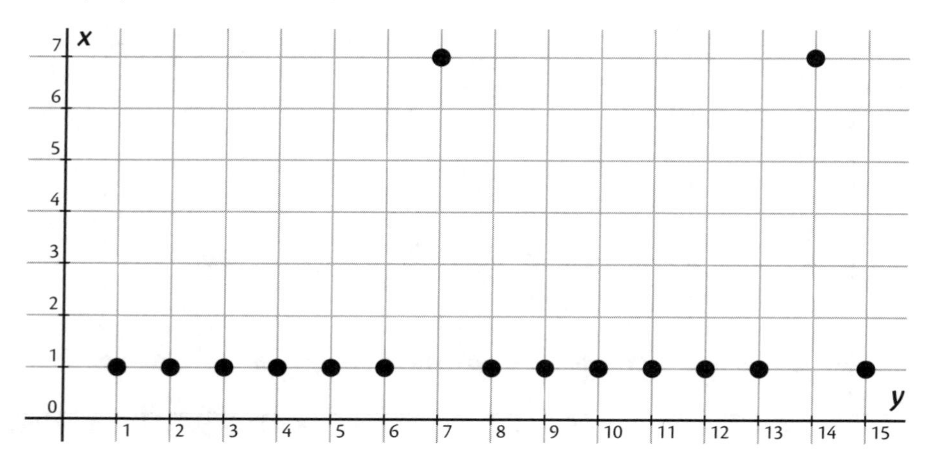


$y$ = GCF(4,$x$)

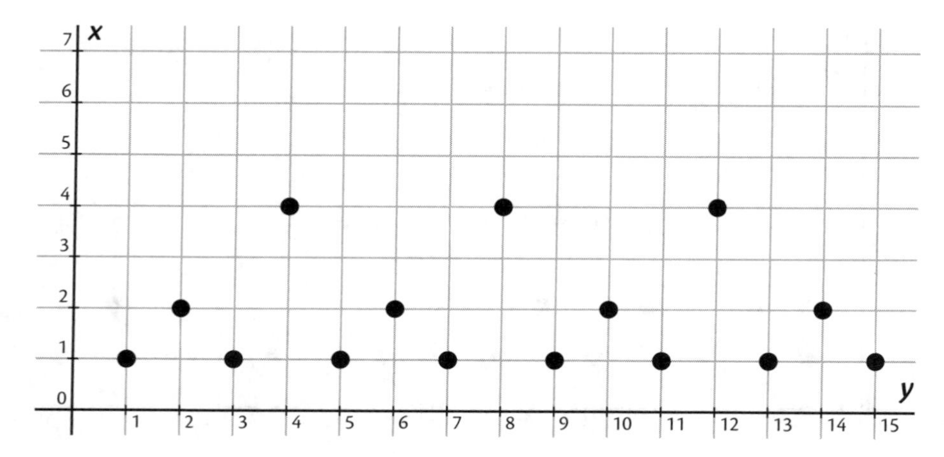


Students should notice that in the first four cases, there are "mountains" at the multiples of the value $a$ in the equation $y$ = GCF($a$,$x$). But in the last example there are two "sizes" of mountains. So students need to then explore what is different between the first four cases and the last.

Ideally, they will realize that when the value of $a$ is a prime number, the graphs will all have the first look, with the mountains at values of $x$ that are multiples of $a$. But when $a$ is not a prime, there are possible factors other than the number itself, so there are different sizes of mountains.

## ASSESSING MATHEMATICAL PRACTICE STANDARD 8

In assessing student proficiency with Mathematical Practice Standard 8, recognizing and using regularity in repeated reasoning, there are a number of things to look for:

- Does the student look for shortcuts for moving from one situation to a similar one?
- Is the student systematic enough to allow him or her to notice patterns or regularities?
- Does the student notice similarities in related situations, formulate conjectures, and test them?

## SUMMARY

In order for students to engage in Mathematical Practice Standard 8:

- Teachers should provide problems that lead to generalizations.
- Teachers should often be asking, *What is happening over and over? What do those values have in common?*

# • CHAPTER 9 •

# Using Visualization and Performing Mental Math and Estimation

**IN ADDITION** to the eight practice standards that U.S. teachers attend to, many Canadian teachers also focus on two additional processes: visualization, and mental math and estimation. U.S. teachers could also benefit from encouraging these processes.

## VISUALIZATION

Visualization involves students visualizing a situation to help them make better mathematical sense of that situation. Although it may seem obvious that visualization is valuable in geometry, it is equally valuable in other content strands, whether number, pattern and algebra, measurement, or data. Visualization might also be viewed as internalized modeling.

### In Number

For example, young students use visualization when subitizing, to decide the amount being shown when the amounts are small. (Subitizing involves the ability to simply see the amount without counting.) Somewhat older students visualize $3 \times 4$ by imagining an array:

v v v v
v v v v
v v v v

In this way, they can see why $3 \times 4 = 4 \times 3$ by turning the image mentally 90° or by indicating that they see 3 sets of 4 when looking at rows, but 4 sets of 3 when looking at columns in the same picture.

Older students might visualize the concept of least common multiple, for example, in this case of 4 and 6, by looking for the first time when a train of 4s matches a train of 6s.

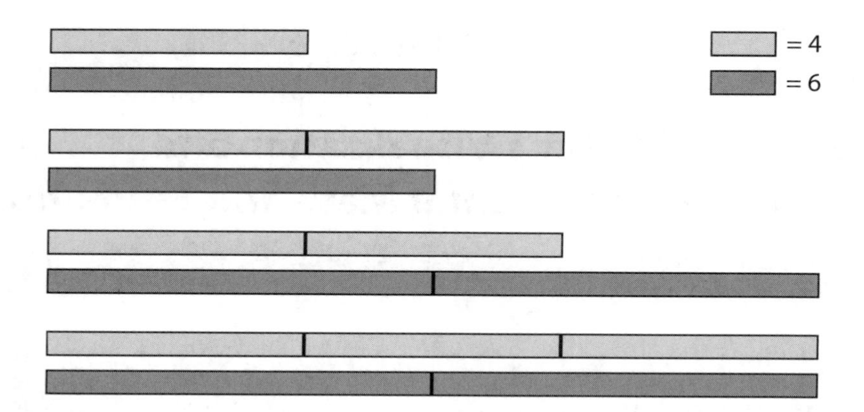


## In Pattern and Algebra

Young students might visualize a pattern like 2, 4, 6, 8, . . . as groups of 1 child's eyes, 2 children's eyes, 3 children's eyes, and so forth. They could visualize equations as pan balance situations. For example, 4 + □ = 8 is about visualizing what to add to a group of 4 to balance 8.

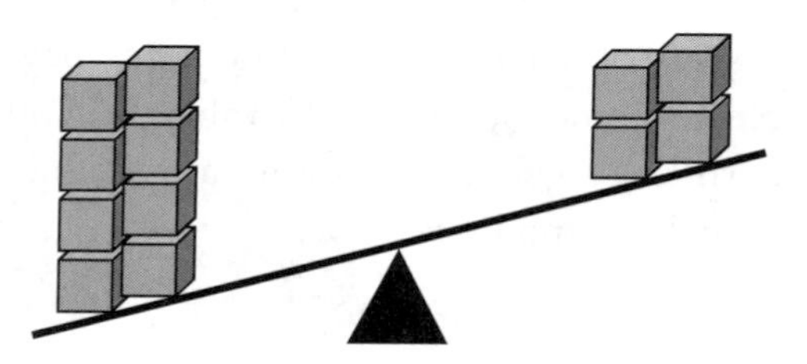

Older children might visualize a pattern like 3, 8, 13, 18, 23, 28, . . . as shown below in order to help them see what the 30th term of that pattern is.

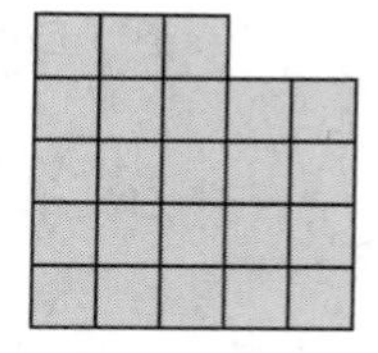

Notice that the 1st term, 3, is the number of squares in the 1st row; the 2nd term, 8, is the number of squares in the first 2 rows; the 3rd term, 13, is the number of squares in the first 3 rows; the 4th term, 18, is the number of squares in the first 4 rows, etc.

Thus it becomes pretty clear that the 5th term is 5 rows of 5, less the 2 squares missing in the first row, so the 30th term will be $30 \times 5 - 2$.

Older students can solve equations by visualizing using algebra tiles. For example, they might visualize $2x + 12 = 4x + 5$ like this:

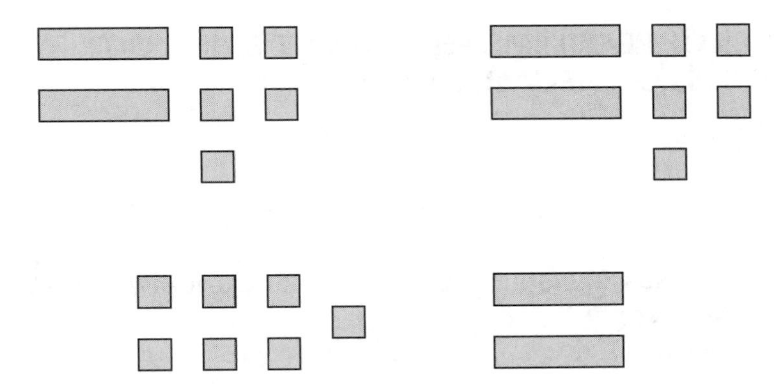

Because the $2x + 5$ on the left matches the $2x + 5$ on the right, it is only comparing 7 to $2x$ that matters. The only way the relationship can be true is if each $x$ bar is worth $3\frac{1}{2}$.

## In Measurement

Young students visualize measurements like 10 inches in terms of familiar benchmarks, such as a ruler of 12 inches.

Somewhat older students visualize perimeter as the "unwinding" of the sides of a shape into a continuous line segment. For example, the perimeter of this square is unwound to be the line segment shown:

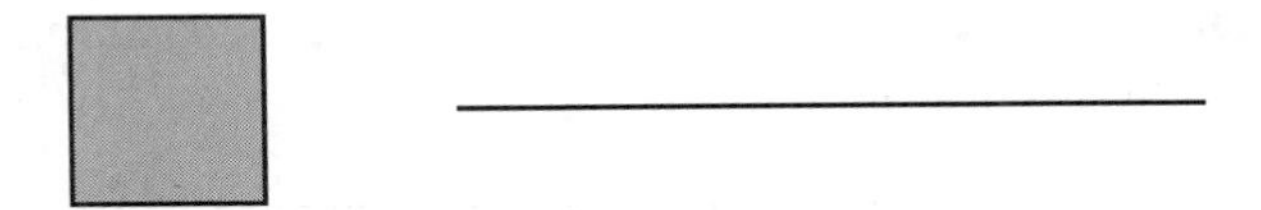

Older students visualize the volume of a prism in terms of the product of its base and height by visualizing the prism as layers of the base.

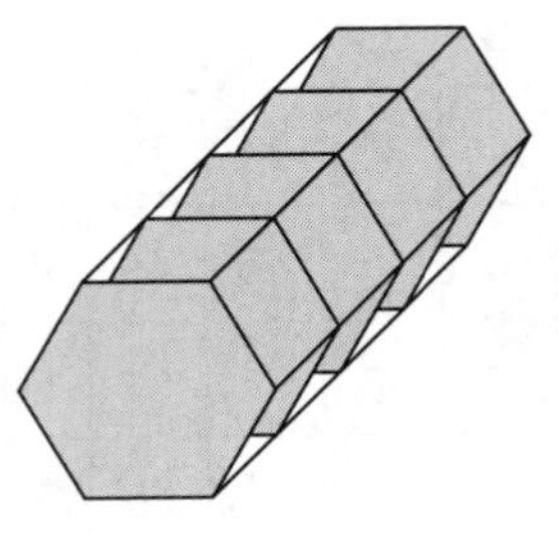

## In Data and Probability

Students of all ages view data visually when they read and interpret various types of graphs.

## EXAMPLES OF PROBLEMS THAT MIGHT BRING OUT THE PROCESS OF VISUALIZATION

### Grades K–2

**8 Dots**

? How would you arrange 8 dots to make it easy to quickly see it as 8?

Asking students to think about what makes an amount easy to see versus less easy to see focuses them on the power of visualization.

It is likely that students will arrange the 8 in two groups of 4, or perhaps a 5 and a 3 on a ten-frame.

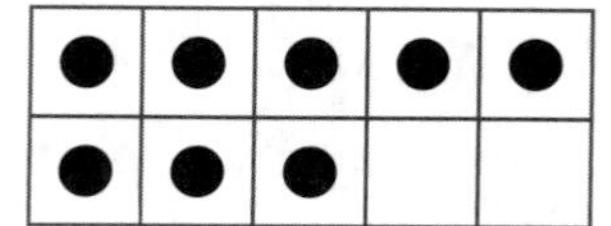

**18 and 82**

? How could you make it easy to see that 18 and 82 together make 100?

It is likely that students will visualize 18 and 82 making 100 by thinking of a hundred-chart. For example:

| 1 | 2 | 3 | 4 | 5 | 6 | 7 | 8 | 9 | 10 |
|---|---|---|---|---|---|---|---|---|---|
| 11 | 12 | 13 | 14 | 15 | 16 | 17 | 18 | 19 | 20 |
| 21 | 22 | 23 | 24 | 25 | 26 | 27 | 28 | 29 | 30 |
| 31 | 32 | 33 | 34 | 35 | 36 | 37 | 38 | 39 | 40 |
| 41 | 42 | 43 | 44 | 45 | 46 | 47 | 48 | 49 | 50 |
| 51 | 52 | 53 | 54 | 55 | 56 | 57 | 58 | 59 | 60 |
| 61 | 62 | 63 | 64 | 65 | 66 | 67 | 68 | 69 | 70 |
| 71 | 72 | 73 | 74 | 75 | 76 | 77 | 78 | 79 | 80 |
| 81 | 82 | 83 | 84 | 85 | 86 | 87 | 88 | 89 | 90 |
| 91 | 92 | 93 | 94 | 95 | 96 | 97 | 98 | 99 | 100 |

But perhaps students will consider a base ten block arrangement or some other visual tool. What is important is that they begin to see useful number "partners," as in the case of 18 and 82 partnering to make 100.

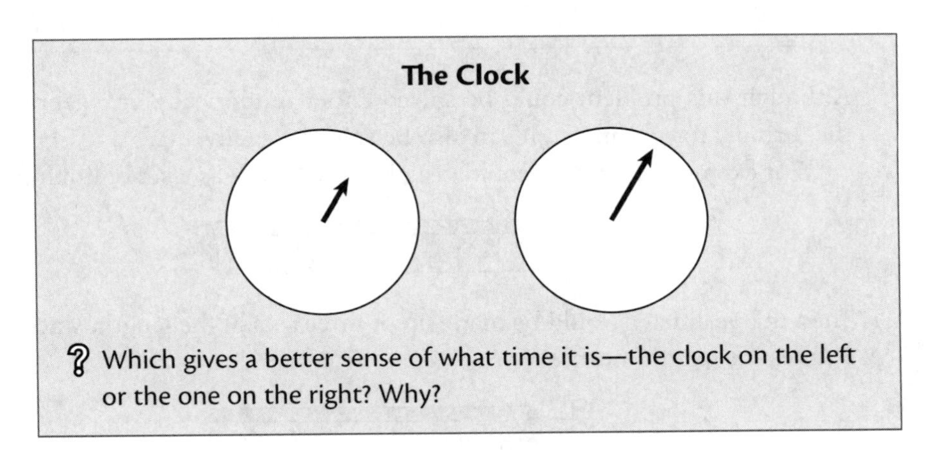


This question focuses students on how much more information the hour hand provides than the minute hand.

A student should observe that the clock on the left shows a time of around 1 o'clock or 2 o'clock or somewhere between the two, whereas the clock on the right indicates that it is about 5 minutes or 6 minutes after the hour, but provides no idea of what part of the day is being represented.

## Grades 3–5

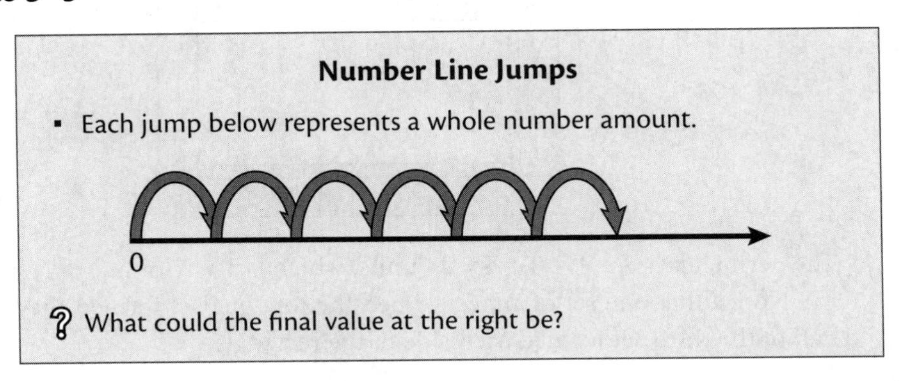


A student responding to the question above will recognize multiplication by 6 as the result of 6 identical actions. This is a good way to visualize multiplication.

Hopefully, the student will realize that the value at the right end of the set of jumps could be any whole number multiple of 6 other than 0 (e.g., 6 or 12 or 60 or 30 or 600).

**Perimeter and Length**

- A rectangle has a perimeter 3 times its length.

? What could its dimensions be?

Although this problem could be solved either numerically or algebraically using the formula for perimeter, it can also be solved visually.

For example, a student could visualize the length as a set of linking cubes:

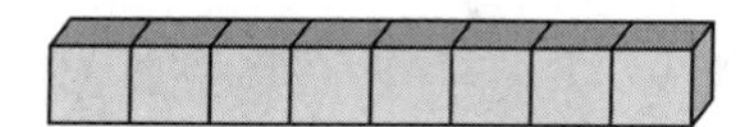

Then the perimeter would be made up of three sets of the length, since that's what 3 times the length means:

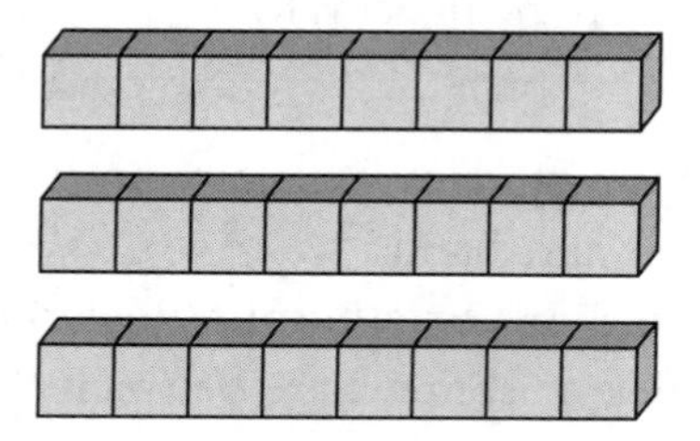

The three sets of cubes could be rearranged like this, forming an $8 \times 4$ rectangle in the interior:

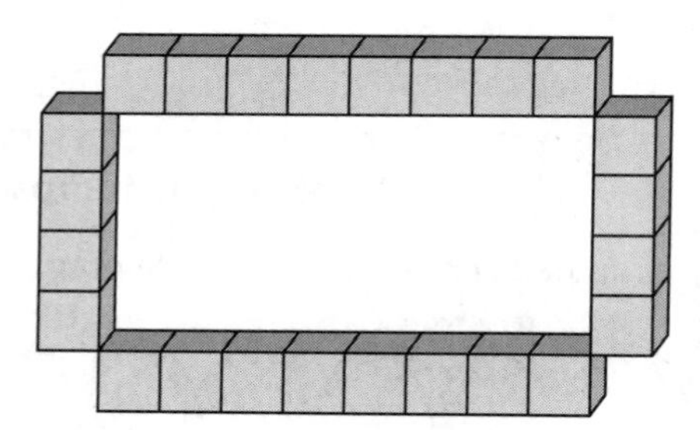

The perimeter is $8 + 4 + 8 + 4 = 24$ units, which is $3 \times 8$ units.

Notice that one set of cubes formed the top, one set formed the bottom, and half of the third set formed each side of the rectangle.

**Perimeter and Area**

- One rectangle has half the area, but almost the same perimeter, as the rectangle it is half of.

? What could the rectangles look like?

A student might visualize that if you keep most of the perimeter of a rectangle when you cut it in half, the rectangle must be fairly thin.

For example, when you cut this square in half, you seem to lose a fair bit of the perimeter:

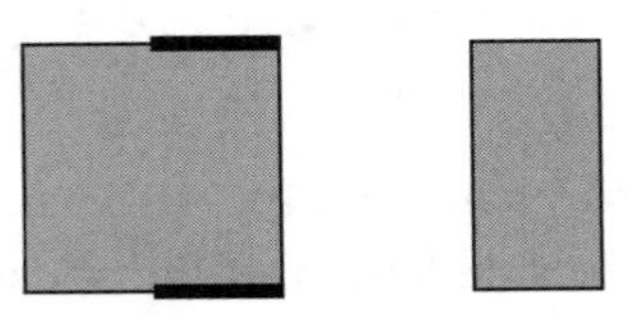

But when you cut a skinnier rectangle in half, you lose very little perimeter if you cut it to be even skinnier:

## Grades 6–8

**Fraction Division**

- One way to view fraction division is to think of unit rate.
- Consider this expression: $\frac{1}{3} \div \frac{2}{5}$

? If you can complete $\frac{1}{3}$ of a job in $\frac{2}{5}$ of an hour, how much can you do in an hour?

A student might use visualization to see how to solve this problem.

For example, the picture below indicates that you can do $\frac{1}{3}$ of a job in $\frac{2}{5}$ of an hour:

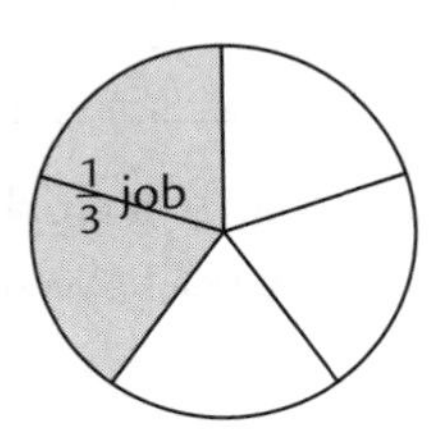

It is easy to see that in each fifth of an hour, you can complete $\frac{1}{3} \div 2 = \frac{1}{6}$ of the job. So in a whole hour, you can complete $\frac{5}{6}$ of the job: $\frac{5}{6}$ is $\frac{1}{3} \times \frac{5}{2} = \frac{1}{3} \div \frac{2}{5}$.

**Area of a Circle**

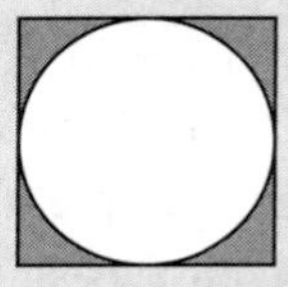

? How does this picture help you see that the area of a circle is less, but not a lot less, than 4 times the square of the radius?

Students can observe that the area of the circle is less than the area of the surrounding square. The side length of the square is $2r$, where $r$ is the radius of the circle, so its area is $4r^2$. Just a quick glance suggests that the circle might be about $\frac{3}{4}$ of the full area, so about $3r^2$.

**Percent**

? How can you visualize 90% of 60?

Ideally, students think of 90% of something as almost all of the something.

Some students might estimate and show a set of 60 items and say 90% is most of them, for example, about 55.

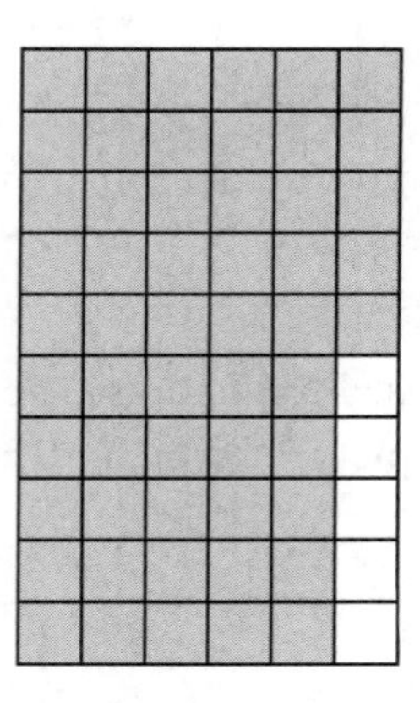

Other students might visualize a clock and realize that 10% of an hour, which is 60 minutes, is 6 minutes, so 90% is 54 minutes.

Still others might visualize a number line that is 60 units long, divide it into 10 equal sections, and keep 9 of them.

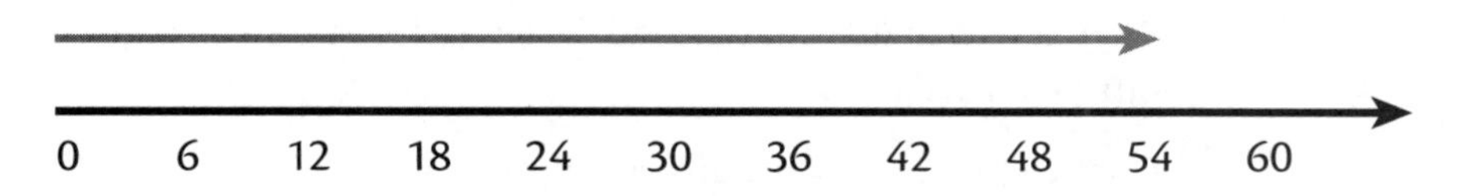


## MENTAL MATH AND ESTIMATION

Mental math is about becoming fluent enough with numbers that paper and pencil algorithms are not required, nor are calculators. Rather, students can use a variety of strategies to flexibly calculate mentally. Mental math is regarded as a process because it is seen as a way of thinking that crosses grade levels.

Getting students to use this process requires, of course, asking them to do so, that is, asking them to calculate mentally and asking them to estimate. But students need to learn to do this, and asking the right kinds of questions is important to make that learning happen.

## EXAMPLES OF PROBLEMS THAT MIGHT BRING OUT THE PROCESS OF MENTAL MATH AND ESTIMATION

### Grades K–2

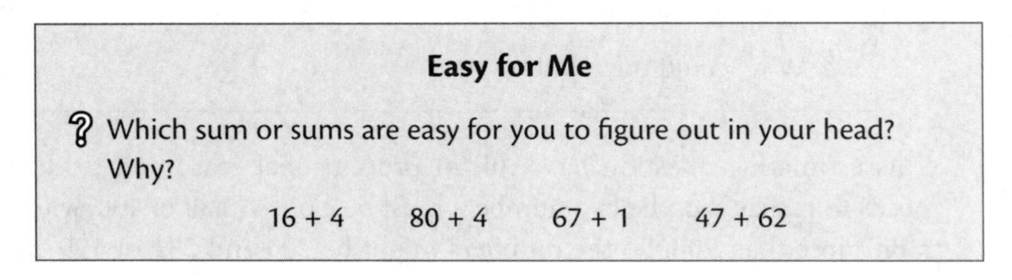

**Easy for Me**

Which sum or sums are easy for you to figure out in your head? Why?

16 + 4 80 + 4 67 + 1 47 + 62

Students might have different opinions on this question. Some might find all of the questions easy.

Others might find only the first one easy because they get nervous with greater values in a problem.

Some students might think that the second sum is easy since it's just another way to say 84.

Some might find the third one easy since the answer is just the next number.

And some might think that the fourth sum is easy because the digits in the ones column total less than 10 (no regrouping necessary).

**Move It Around**

- Alanna said that to add 87 and 48, a good idea is to change the question to 85 + 50.

? Does this give the same answer? Why or why not?

How else might you change 87 + 48 to make it easy to figure out in your head?

This question is useful because it is a valuable application of the associative property of addition. The sum 87 + 48 is actually (85 + 2) + 48, which has the same answer as 85 + (48 + 2).

Students might explain this without actually referring to the property, by suggesting something like this: *If you had a pile of 87 and a pile of 48, you could move 2 from the 87 pile over to the 48 pile and you would still have the same amount.*

Another useful mental strategy is to think of 87 + 48 as 90 + 45, or perhaps 100 + 35.

**About 400**

- You add two numbers that are fairly close together, and the answer is just a little more than 400.

? What could the numbers be?

This estimating question has a bit of proportional reasoning to it; the student needs to realize that the two numbers must be close to half of 400, with at least one a bit more than 200. So the numbers might be 200 and 201 or 199 and 203, or a similar pairing.

## Grades 3–5

### Half and Double

? When might it be useful to multiply two numbers by taking half of one and doubling the other?

If you were multiplying an even number by 5, you might want to take half of that even number and multiply by 10 instead, since it's easy to multiply by 10.

Or if you were multiplying an even number by 50, you might want to take half of that even number and multiply by 100 instead, since it's easy to multiply by 100.

### Dividing in Your Head

? Which of these calculations would you find easy to do in your head? Why those?

$142 \div 5 \qquad 297 \div 3 \qquad 1248 \div 4 \qquad 1448 \div 8$

Some students will find all of these calculations easy; for others, only some of them will be easy.

The first division might be easy if it were transformed into $284 \div 10$, with a result of 28.4. It is easier now because dividing by 10 does not require many steps. The values are the same since the amount each person gets if 142 is shared by 5 is the same as the amount each person gets if twice as much, 284, is shared by 10.

The second might be easy for a student who thinks of 297 as $300 - 3$. There are 100 groups of 3 in 300, so there is one less group of 3 in 297.

The third might be easy for a student who thinks of 1248 as $1200 + 48$. Each part can easily be divided by 4 to get $300 + 12$, so the result is 312.

The fourth might be easy for someone who thinks to repeatedly take half, that is, $1448 \div 8 = 724 \div 4 = 362 \div 2$, which is 181.

### About $1\frac{3}{4}$

- You have added two fractions. Both are proper fractions.
- Their sum is just a little less than $1\frac{3}{4}$.

? What could the fractions be?

Students should be able to address this question using mental math. For example, a student might realize that $1\frac{3}{4} = 1 + \frac{3}{4}$. So a sum like $\frac{9}{10} + \frac{3}{4}$ would be a bit less than $1 + \frac{3}{4}$; it fits the bill since $\frac{9}{10}$ and $\frac{3}{4}$ are both proper fractions.

Alternatively, a student could think of $1\frac{3}{4}$ as $\frac{7}{4}$, which is equivalent to $\frac{14}{8}$. A number that is a bit less than $\frac{14}{8}$ is $\frac{14}{9}$. Two proper fractions that sum to $\frac{14}{9}$ are $\frac{8}{9}$ and $\frac{6}{9}$, so they would work as answers to the problem.

## Grades 6–8

### Calculating Percents

? Which of these calculations would be easy to do in your head? Explain why.

90% of 180 5% of 412 99% of 412 15% of 380

Some students might decide that all of these calculations are easy to do in their head.

For example, the first calculation could be done by subtracting 10% of 180 (or 18) from 180, which gives 162.

The second might be considered easy if it is transformed from 5% of 412 to 10% of 206, or 20.6.

The third might be considered easy if it is viewed as 1% of 412 less than 412, or 412 – 4.12, or 407.88.

The fourth might be considered easy if 15% is calculated as 10% plus another half of that. So 15% of 380 would be 38 + 19 = 57.

### Dividing Rationals

- You divide two rational numbers and the quotient is about $-3\frac{1}{3}$.

? What could the rational numbers be?

Students should be able to recognize that one number must be positive and one negative. They should also realize that the absolute value of one of the amounts should be a bit more than 3 times the absolute value of the other amount. So, for example, the answer could be –10 and 3, or it could be 5 and $-\frac{3}{2}$, or it could be –5 and $\frac{5}{3}$.

**Square Roots**

What would be your estimates (without using a calculator) for each of these square roots? Why?

$\sqrt{20}$ $\sqrt{178}$ $\sqrt{389}$ $\sqrt{1200}$

Students should be able to estimate all of these. For example, the first is probably about halfway between 4 and 5, the second is maybe 13.2 or 13.3 since $13 \times 13 = 169$, the third about 19 since $20 \times 20 = 400$, and the last about 35 since $30 \times 30 = 900$ and $40 \times 40 = 1600$.

## ASSESSING VISUALIZATION AND MENTAL MATH AND ESTIMATION

In assessing student proficiency with visualization and mental math and estimation, there are a number of things to look for:

- Does the student regularly use manipulatives or sketch a picture to show the broad strokes of a problem?
- Can the student interpret a visual of a mathematical situation effectively?
- Does the student relate a complex calculation to ones that are simpler to do mentally?
- Does the student regularly use estimation to check the reasonableness of his or her computational solutions?
- Does the student know when estimation is sufficient and when it is not?

## SUMMARY

In order for students to engage in visualization and mental math and estimation:

- Teachers should regularly encourage students to draw pictures to represent problems. These pictures need not be—and perhaps *should* not be—accurate pictures of the objects being represented, but simply mathematical representations of them.
- Teachers should not only ask students to use estimation and mental math but also provide opportunities for students to consider which calculations really are easy to do with mental math. Teachers should also ask for estimation, without necessarily using the word "estimate," for example, asking for two numbers that have a product that is "about" a certain amount.

• • •

# Conclusion

**THE EXAMPLES** in this book are just that—examples. There are so many more possibilities for bringing out the important mathematical processes. My message is that it is problems that *explicitly* evoke standards of practice that are the ones that may make the most difference in helping students think like young mathematicians. It is also helpful for students if teachers incorporate terms such as "model" and "structure" and "mathematical tools" appropriately in their instruction so that students understand when and how they are engaging in these practices.

Even though each problem appeared in only one chapter, it is often the case that a number of practices might be associated with a single problem (as was frequently indicated). Instilling these practices in students is not so much about meeting a checklist of how many practices are being covered in a particular day; it is more a matter of focusing teaching on these kinds of problems. The rest will then take care of itself.

• • •

# Bibliography


Common Core State Standards Initiative. (2010). *Common Core State Standards for Mathematics.* Available at http://www.corestandards.org/wp-content/uploads/Math_Standards1.pdf

Dole, S., Clarke, D., Wright, T., & Hilton, G. (2010). Students' proportional reasoning in mathematics and science. Available at http://www.proportionalreasoning.com/uploads/1/1/9/7/11976360/students_proportional_reasoning_in_mathematics_and_science.pdf

Duckworth, A. L., Peterson, C., Matthews, M. D., & Kelly, D. R. (2007). Grit: Perseverance and passion for long-term goals. *Journal of Personality and Social Psychology, 92,* 1087–1101.

Grimes, J. (2012). *43 McNuggets* [video]. Available at https://www.youtube.com/watch?v=vNTSugyS038/

Mateas, V. (2016). Debunking myths about the Standards for Mathematical Practice. *Mathematics Teaching in the Middle School, 22,* 93–99.

Ministère de l'Éducation, Gouvernement du Québec. (2001). *Québec Education Program.* Available at www.education.gouv.qc.ca

Ministry of Education Ontario. (2005). *The Ontario curriculum Grades 1–8: Mathematics.* Available at www.edu.gov.on.ca

National Council of Teachers of Mathematics (NCTM). (1989). *Curriculum and evaluation standards.* Reston, VA: Author.

National Council of Teachers of Mathematics (NCTM). (2000). *Principles and standards for school mathematics.* Reston, VA: Author.

Polya, G. (1957). *How to solve it* (2nd ed.). Princeton, NJ: Princeton University Press.

Small, M. (2012). *Good questions: Great ways to differentiate mathematics instruction* (2nd ed.). New York, NY: Teachers College Press.

Small, M. (2017). *Making math meaningful to Canadian students, K–8* (3rd ed.). Toronto: Nelson Education.

Western and Northern Canadian Protocol. (2006). *The common curriculum framework for K–9 mathematics.* Available at www.wncp.ca

Zollman, A. (2009). Mathematical graphic organizers. *Teaching Children Mathematics, 16,* 222–231.

# Index

## INDEX OF SUBJECTS AND CITED AUTHORS

# INDEX OF STANDARDS FOR MATHEMATICAL PRACTICES AND PROCESSES

## CANADIAN CURRICULA PROCESS STANDARDS



## COMMON CORE STANDARDS FOR MATHEMATICAL PRACTICE

## NATIONAL COUNCIL OF TEACHERS OF MATHEMATICS (NCTM) PROCESS STANDARDS

• • •

# About the Author

**MARIAN SMALL** is the former dean of education at the University of New Brunswick. She speaks regularly about K–12 mathematics instruction.

She has been an author on many mathematics text series at both the elementary and the secondary levels. She has served on the author team for the National Council of Teachers of Mathematics (NCTM) Navigation series (pre-K–2), as the NCTM representative on the Mathcounts question writing committee for middle school mathematics competitions throughout the United States, and as a member of the editorial panel for the NCTM 2011 yearbook on motivation and disposition.

Dr. Small is probably best known for her books *Good Questions: Great Ways to Differentiate Mathematics Instruction* and *More Good Questions: Great Ways to Differentiate Secondary Mathematics Instruction* (with Amy Lin). *Eyes on Math: A Visual Approach to Teaching Math Concepts* was published in 2013, as was *Uncomplicating Fractions to Meet Common Core Standards in Math, K–7.* In 2014, she authored *Uncomplicating Algebra to Meet Common Core Standards in Math, K–8.* She is also author of the three editions of a text for university preservice teachers and practicing teachers, *Making Math Meaningful to Canadian Students: K–8*, as well as the professional resources *Big Ideas from Dr. Small: Grades 4–8*; *Big Ideas from Dr. Small: Grades K–3*; and *Leaps and Bounds toward Math Understanding: Grades 3–4*, *Grades 5–6*, and *Grades 7–8*, all published by Nelson Education Ltd. More recently, she has authored a number of additional resources focused on open questions, including *Open Questions for the Three-Part Lesson, K–3: Numeration and Number Sense*; *Open Questions for the Three-Part Lesson, K–3: Measurement and Pattern and Algebra*; *Open Questions for the Three-Part Lesson, 4–8: Numeration and Number Sense*; and *Open Questions for the Three-Part Lesson, 4–8: Measurement and Pattern and Algebra*, all published by Rubicon Publishing.

She also led the research resulting in the creation of maps describing student mathematical development in each of the five NCTM mathematical strands for the K–8 levels and has created the associated professional development program, PRIME. She is currently working with Rubicon Publishing to develop a K–8 digital teaching resource, *MathUp*, that assists teachers to deliver a math program that prioritizes the development of students as mathematical thinkers.